HOUSES

VO TRONG NGHIA

&

THE WORK OF VTN ARCHITECTS

HOUSES

VO TRONG NGHIA &
THE WORK OF VTN ARCHITECTS

Preface by VTN Architects
Introduction by Vo Trong Nghia
Edited by Oscar Riera Ojeda

OSCAR RIERA OJEDA
PUBLISHERS

Contents

Preface

VTN Architects

In an age of overpopulation and rapid urban development, most Vietnamese houses are located within crowded cities. Therefore, the residents suffer from polluted air with little green space for themselves. The presence of televisions, Internet, and social media also poses problems for urban dwellers. People are mentally stressed from constant social pressure and the modern lifestyle. They have limited space for activities, for living life, and for enjoying nature.

Understanding the situation, VTN Architects' "Houses for Trees" connect people to nature, creating pockets of greenery within our concrete cities. Our houses become ecosystems that merge natural energy and natural materials to produce net-zero buildings for society. Water is recycled. Spaces are bathed in natural light and fresh air. Solar panels are installed. Our houses also become hotspots for activities such as planting trees and urban farming. They boost interaction between family members, strengthen the atmosphere within each household, and improve people's mind and body. Our "Houses for Trees" also prevent flooding in the cities, battling global climate change.

The office staff enjoy a working environment of innovation, honesty, and clear thinking by practising meditation daily and maintaining the 5 Precepts. This attitude influences the quality of the projects they create. Ultimately, VTN Architects intend to enhance the lives of others by introducing a sense of well-being and calmness that approaches the spiritual.

Houses for Trees

Introduction

Vo Trong Nghia

Vietnam has enjoyed rapid economic development in recent years, with annual GDP growth of 7% in 2019. Consequently, Vietnamese cities are facing the same problems as many developing metropolises in the world: pollution, overpopulation, high urban density with a severe lack of greenery. The country's population is approximately 97.4 million, mostly in sprawling urban centers. Particularly, Ho Chi Minh City has a population of 8.6 million and a population density of 4,292 inhabitants per square kilometer.

The quality of urban life in Ho Chi Minh city is poor due to the lack of parks and nature, given that there is only 0.7 square meters of green area per resident. We can also precisely measure the unhealthy air quality caused by economic activities and heavy traffic. Despite enjoying massive urban development, they are disconnecting urban dwellers to nature.

At VTN Architects, we aim to provide a solution through new architecture that harmonises with nature by designing "Houses for Trees." For every 100 square meters we take from the cities, we introduce greenery. The more we build, the more we bring nature back to urban living, the happier people are, and the greener our society becomes. Each house that we construct becomes a pocket of parks for the cities, reconciling nature to modern urban dwellers. Being in our houses equals being in a park or a forest within our concrete busy cities. We wrap nature around the lives of our residents to improve Vietnamese high-density living conditions. Nature plays a huge part in our projects, as a means to combat urban problems.

Rejecting the idea that "green architecture" is merely planting trees on buildings, our architecture provides an ecosystem for people to connect with nature. We merge natural energy and natural materials to produce net-zero buildings for society. Therefore, trees are used not as decoration but as an essential component of our houses. They interact with the buildings to limit water and flooding, to filter light and sounds. We recycle water. We bathe our spaces with natural light and fresh air provided by natural ventilation. Hence, our houses do not require air-conditioning and have minimal energy consumption. We apply solar panels on our roofs with a reasonable price. Even though our houses are artificial creations, they become an ecosystem, welcoming birds, insects, and animals to live side by side with humans.

Furthermore, technological development with the rising popularity of television, Internet, and social media have isolated people. As a result, people tend to spend more time indoors, while having limited space for playing sports, or enjoying nature. Our bodies and minds deteriorate through physical inactivity and mental stress due to the constant pressure of society. This modern lifestyle makes us forget how to live well.

However, nature can offer a way to deal with those social issues. Our houses aim to improve people's bodies and minds by introducing activities inside the houses, which in turn connect family members. We provide chances for our residents to maintain and grow plants or to participate in urban farming. Our trees are planted independently from our houses to simplify maintenance so that the residents can take care of plants as a hobby. Moreover, urban farming allows family members to come together and grow vegetables in their own houses. These activities boost the interaction among members and improve the family atmosphere within each household. Through bringing nature into daily activities, our houses improve people's health and bring people together regardless of mental stress, body inactivity, and social isolation.

The decision to emphasize planting trees also comes from Vietnam's climate. Plants grow well under tropical conditions with no harsh winter, high temperature, high humidity, and high precipitation. However, it also comes with heavy rains, and flooding poses a problem for the city streets. Our houses prevent flooding by letting rainwater be absorbed into plants instead of leaving it to flow directly into concrete and the city. The more plants we grow in our houses, the more rainwater we absorb, and the more we help our cities battling climate change.

Our houses are pioneers in how to bring greenery back to our cities to combat modern urban problems, social issues, and climate change. We hope other architects reference our architecture as approaches to improve people's living conditions in modern cities and to better society.

VTN Architects uses the Five precepts of Buddhist teaching as the company foundation:

- *No killing*
- *No stealing*
- *No adultery*
- *No lying*
- *No alcohol or drugs*

Following the precepts allows our staff to build trust amongst each other and with our clients. Furthermore, our staff meditates two hours a day to reduce stress, to enhance our concentration, our ethical and our aesthetic sensitivity. We believe that five-precept maintenance and meditation indeed help to better the society. Ultimately, the office's philosophy is closely tied to our architecture as we seek to improve people's lives by introducing a sense of well-being and calm in our houses that approaches the spiritual.

Opposite page: Stepping Park Houses

Stacking Green

Numerous developing Asian cities are becoming indistinguishable. They are losing their regional characteristics under the influence of furious urban sprawl and commercialism. The population boom further worsens the quality of life as well as the quality of greenery in those developing cities. Ho Chi Minh City, Vietnam's largest metropolitan area, is not an exception. Built in 2011, the private house "Stacking Green" takes an experimental approach in designing to challenge this situation.

Even in advanced nations, people have a desire for more trees in both public and private spaces. So locals are already proactively decorating their streets with plants and flowers to turn their surroundings into a greenwood. "Stacking Green" is driven by the same desire. Instead of limiting plants to the street level, planters are distributed vertically over the façade.

The private house, designed for a 30-year-old couple and their mother, is a typical tube house constructed on a 4 meter wide, 20 meter deep plot. The front and back façades are both composed of layers of concrete planters cantilevered from two sidewalls. Not only does the green façade provide visual comfort for the occupants, but it protects its inhabitants from direct sunlight, street noise, and pollution. Post-calculation for determining passive cooling in the house has shown wind flowing through the porous façade and two skylights, which helps the inhabitants feel comfortable. Because they rarely use air conditioning, electricity only costs 25 USD per month on average. The cost of water can be excluded from the ancillary expenses because plants are watered automatically with a rainwater collecting system.

"Stacking Green" has achieved several international recognition, among them a place in the Top 10 of Most Popular Projects by ArchDaily in 2012.

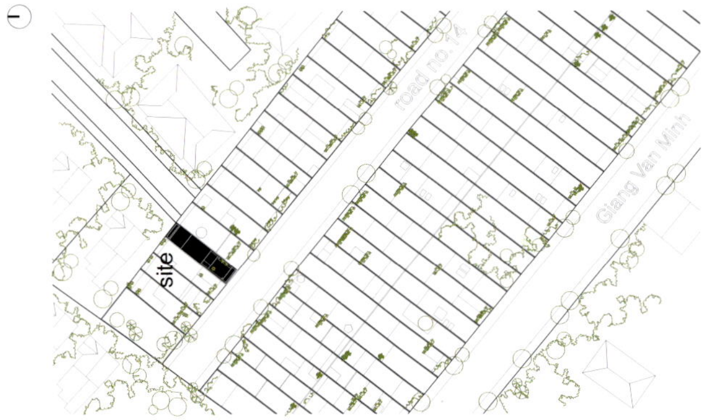

Site

Elevation Sketch

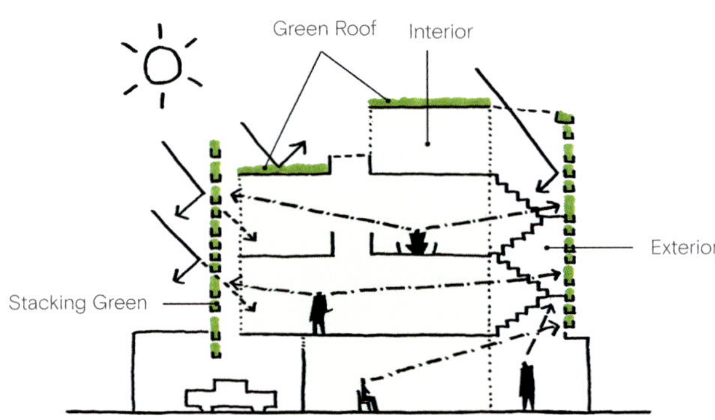

The people can feel the greenery illuminated
by sunlight from everywhere in this house

Section

1 Parking
2 Bedroom 1
3 Courtyard
4 Storage
5 Living Space
6 Kitchen Space
7 Bedroom 2
8 Bathroom
9 Study Room
10 Worship Room
11 Roof Garden

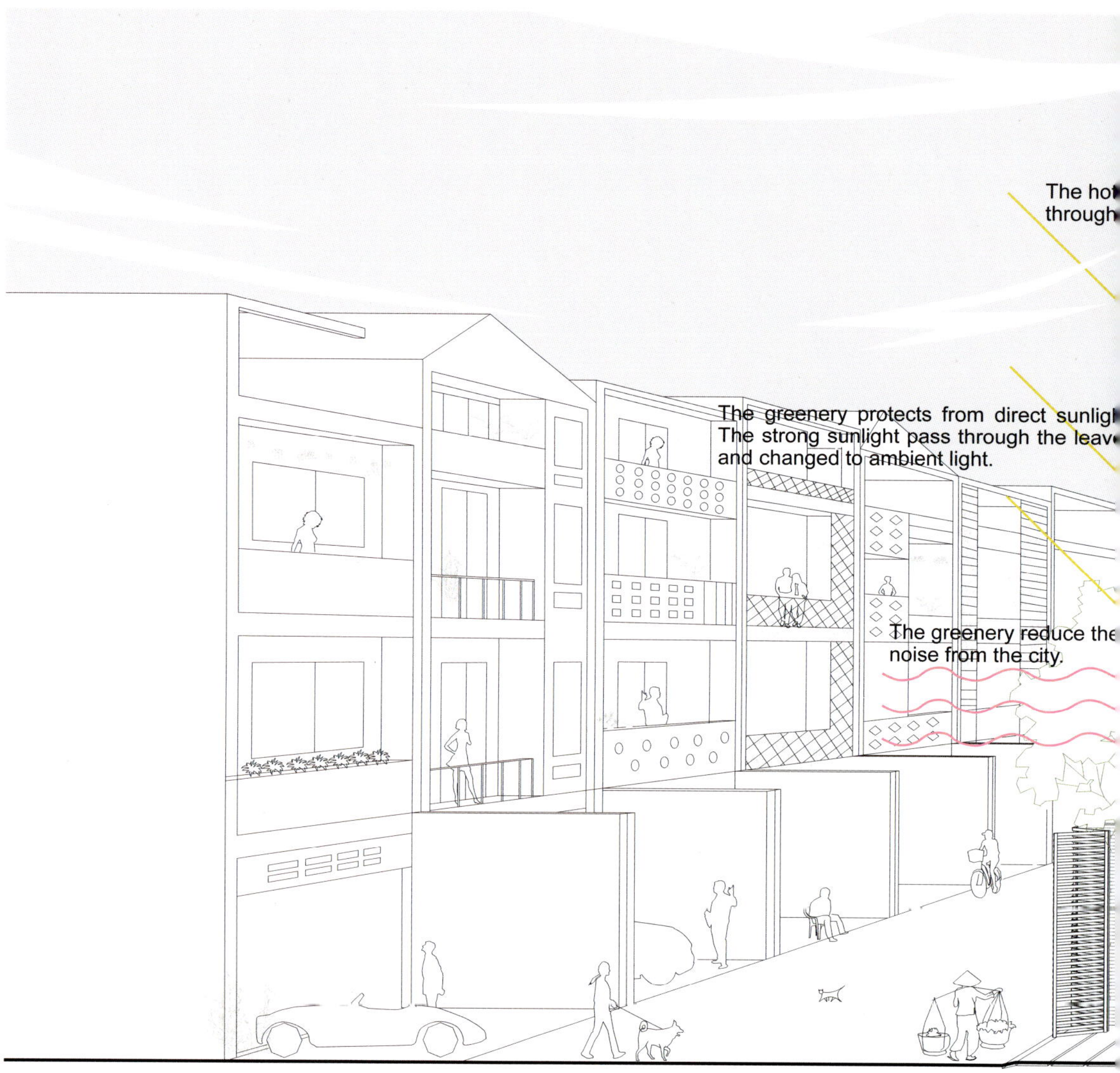

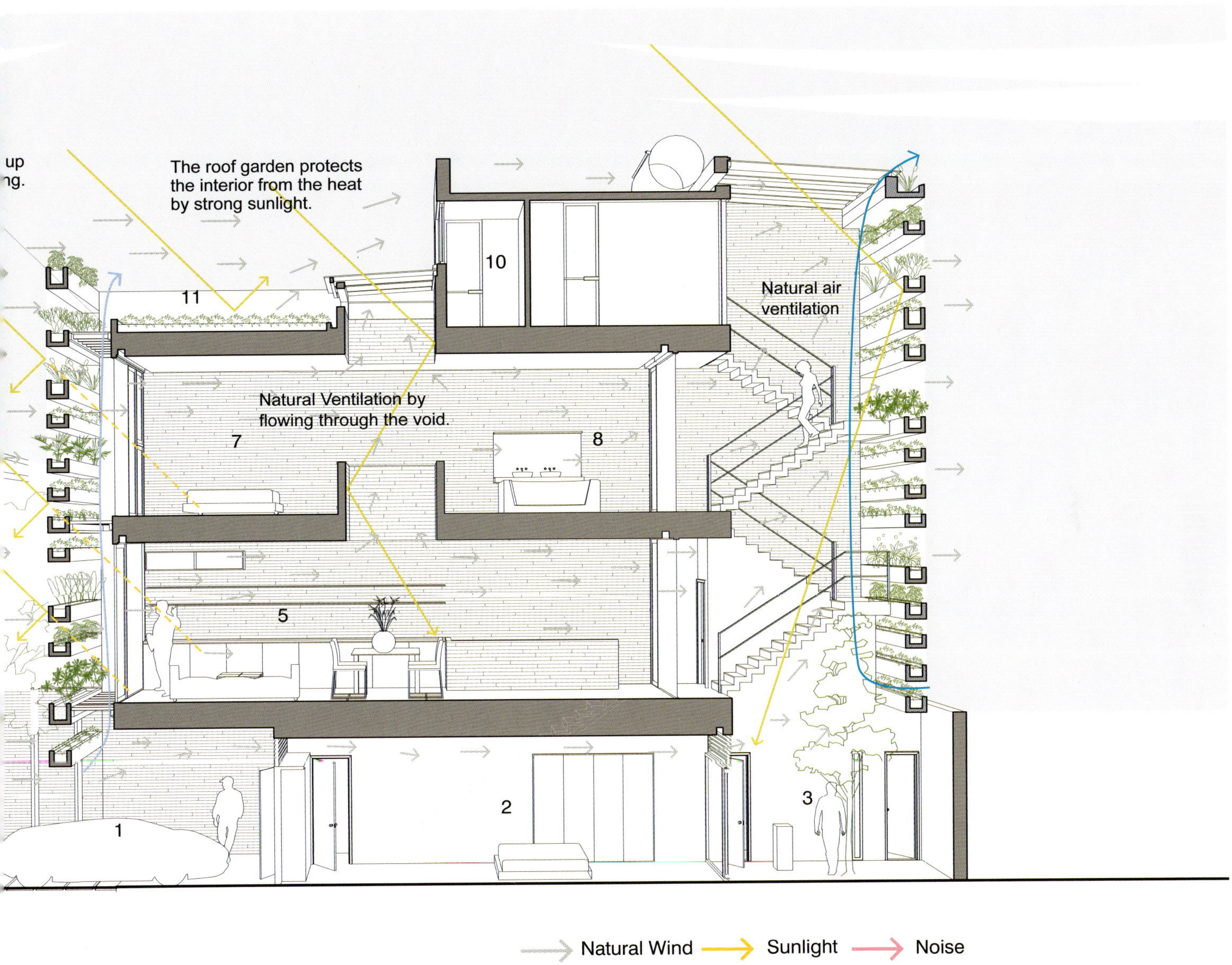
up
ng.
The roof garden protects the interior from the heat by strong sunlight.
10
11
Natural air ventilation
Natural Ventilation by flowing through the void.
7
8
5
2
3
1
Natural Wind
Sunlight
Noise

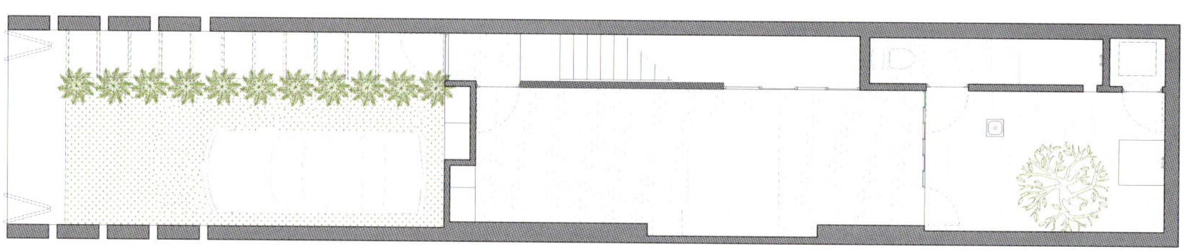

Ground Floor

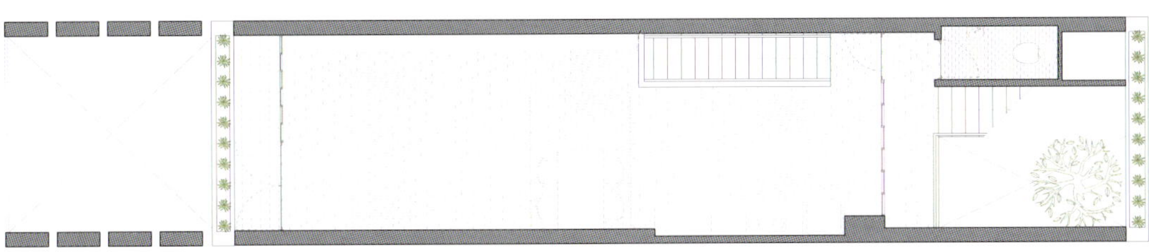

First Floor

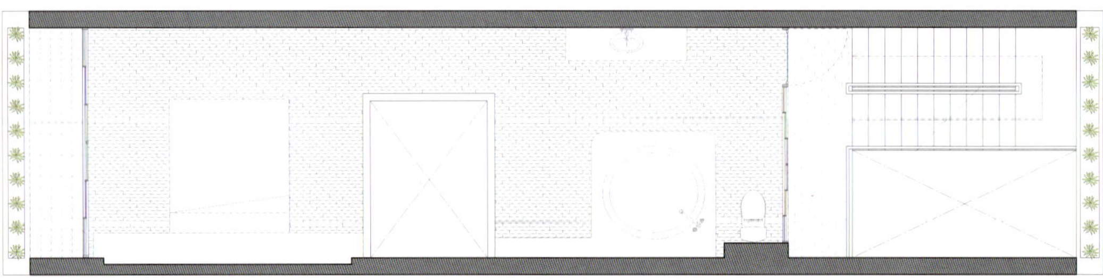

Third Floor

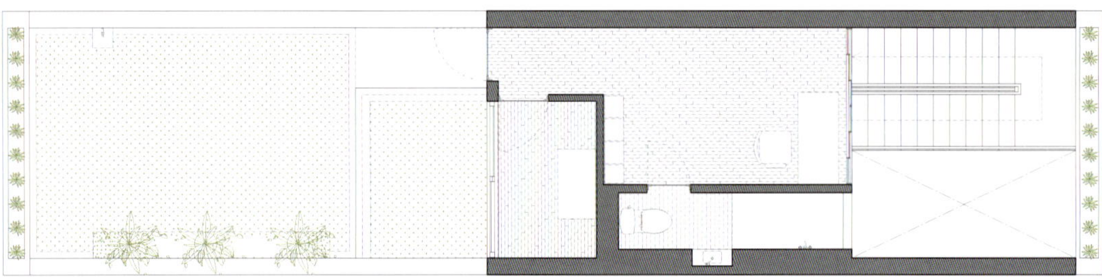

Fourth Floor

Gia Lai House

The Gia Lai House project is located in central Vietnam, where there is a hot and humid climate all year round with an average temperature of 25 degrees Celsius. The long and narrow plot of 7 meters wide and 40 meters deep made for a challenging implementation of the tube house.. As is the case with every other tube house, housing planning is strongly influenced by the layout of the room and the hallway because lots of rooms are needed in a narrow plot. Long corridors become a boring and simple element, the function of space is almost limited.

In this house, VTN Architects have turned the corridor into a continuous and sequential space to serve the functions of living space, dining, and outdoor activities. The width and height change flexibly to bring different experiences between living spaces. Especially in the courtyard, the triangular green tree area creates an impressive feeling for the entire building space. It is a way of breaking down the weak point of the long hallway in the tube house.

The outdoor living space is also another special factor that VTN Architects has applied with Gia Lai House, combined with the skylight and louvers with the theory that to create more light inside the house will reduce the heat from the light and encourage natural ventilation with a chimney effect to provide fresh air to the surrounding rooms.

The wall of the continuous space was finished by 4-meter-long granite stones, which were quarried out in Gia Lai city, functioning as a reflector of the light from the side and above.

Roof Floor

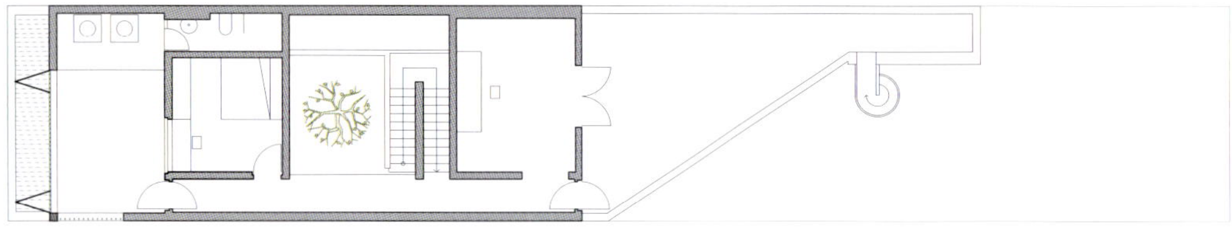

Second Floor

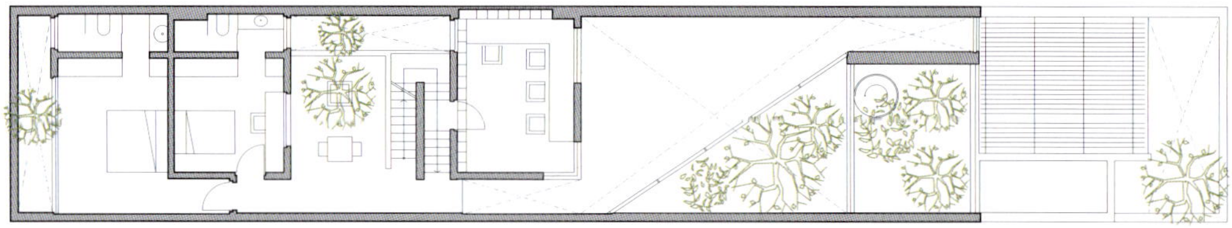

First Floor

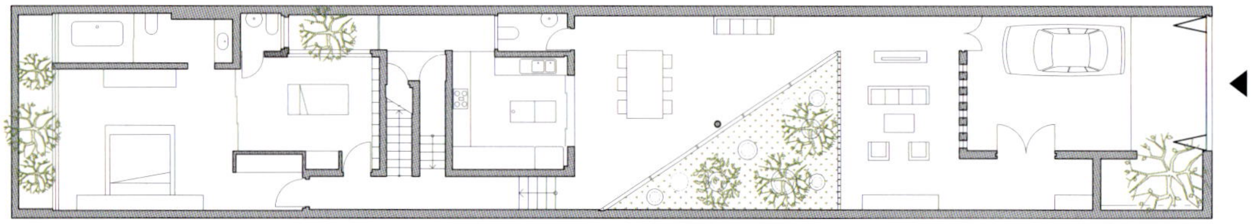

Ground Floor

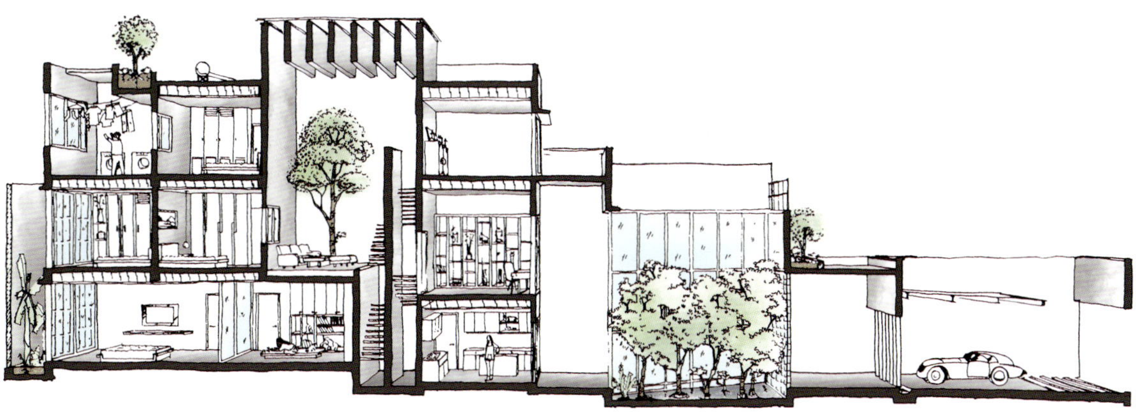

Section Sketch

Section Sketch

Site Plan

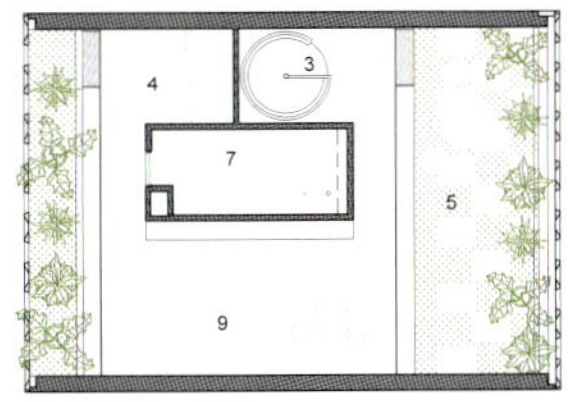

Fifth Floor

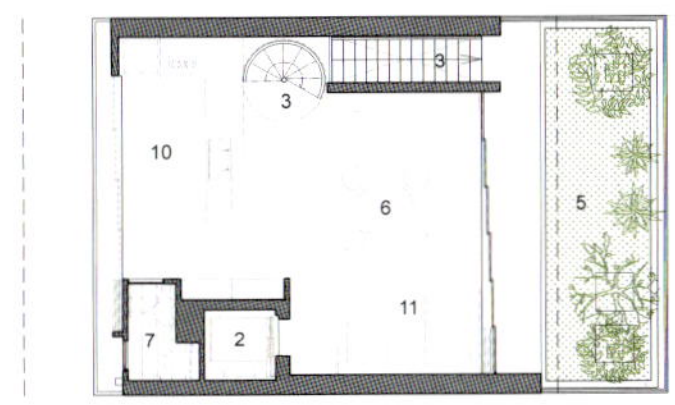

Fourth Floor

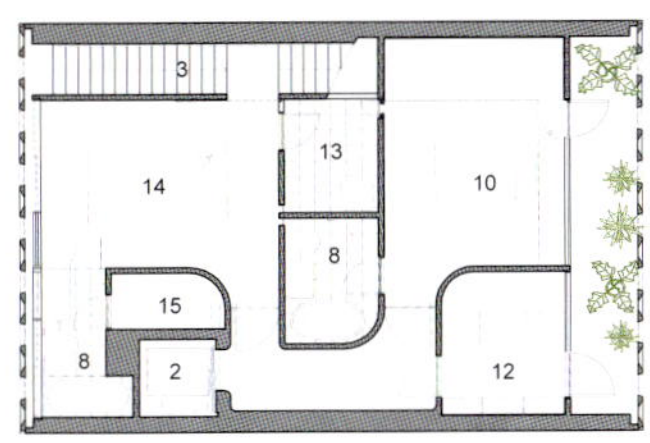

Third Floor

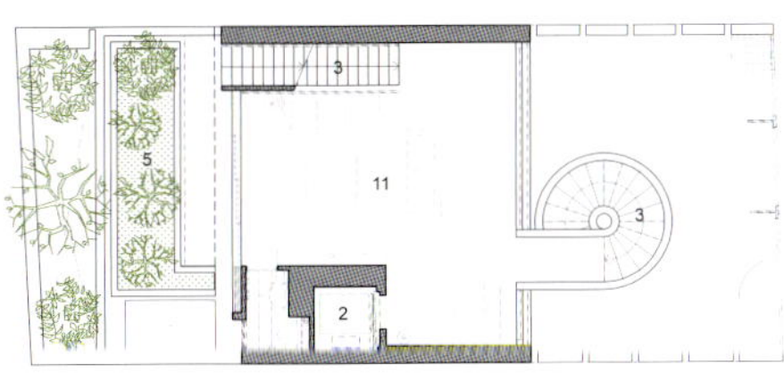

Second Floor

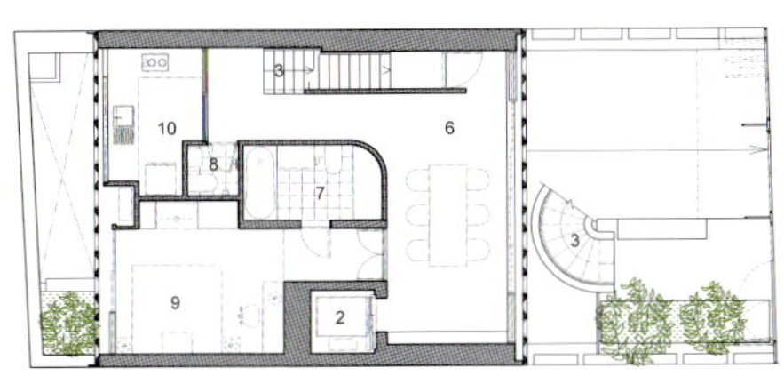

First Floor

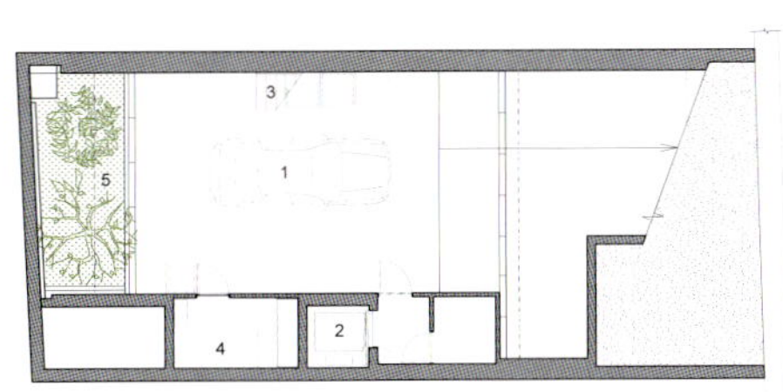

Basement Floor

House for Trees

Under rapid urbanization, cities in Vietnam have diverged far away from their origins as rampant tropical forests. In Ho Chi Minh City, for example, greenery only covers around 0.25% of the city's area. Consequently, the new generations in these developing districts are losing their connections with nature. Instead, they have to deal with serious air pollution and daily traffic congestion caused by an overabundance of motorbikes. These problems caused by too rapid development put many urban citizens under stress.

Within a tight budget of 155,000 USD, "House for Trees" was designed in an effort to change the situation. The project's main idea was to return green spaces to the city, accommodating high-density dwelling with big tropical trees, and use nature as a means to reconnect people with each other as well as to relieve stress.

Five concrete boxes, each a different layout, were designed as "pots" to plant trees on the top. They are positioned to create a central courtyard and small gardens in between. The boxes offer open views to the courtyard with large glass doors and openable windows for natural lighting and ventilation. This central space, shaded by tree spans, also becomes a part of the living space and offers a tropical lifestyle to coexist with nature. Common programs such as a dining room and a library are located on the ground floor, while private bedrooms and a bathroom are on upper floors and connected through the bridges-cums-eaves made of steel. The fact that the living areas are separated requires the homeowner to get out from the boxes more often and thus get closer to nature as well as being more active.

With the concept of a city filled with "House for Trees," the thick soil layer of these "pots" would function as storm-water basins for detention and retention, which would help prevent flooding, a major issue that many cites of Vietnam are facing.

Sketch

Section
1. Approach 2. Storage 3. Front courtyard 4. Central courtyard 5. Library 6. Bathroom
7. Dining room 8. Kitchen 9. Worship room 10. Corridor 11. Bedroom 12. Bridge

Roof floor plan
1. Approach 2. Storage 3. Front courtyard 4. Central courtyard 5. Library 6. Bathroom
7. Dining room 8. Kitchen 9. Worship room 10. Corridor 11. Bedroom 12. Bridge

1st floor plan

1. Approach 2. Storage 3. Front courtyard 4. Central courtyard 5. Library 6. Bathroom 7. Dining room 8. Kitchen 9. Worship room 10. Corridor 11. Bedroom 12. Bridge

2nd floor plan
1. Approach 2. Storage 3. Front courtyard 4. Central courtyard 5. Library 6. Bathroom
7. Dining room 8. Kitchen 9. Worship room 10. Corridor 11. Bedroom 12. Bridge

Section Sketch

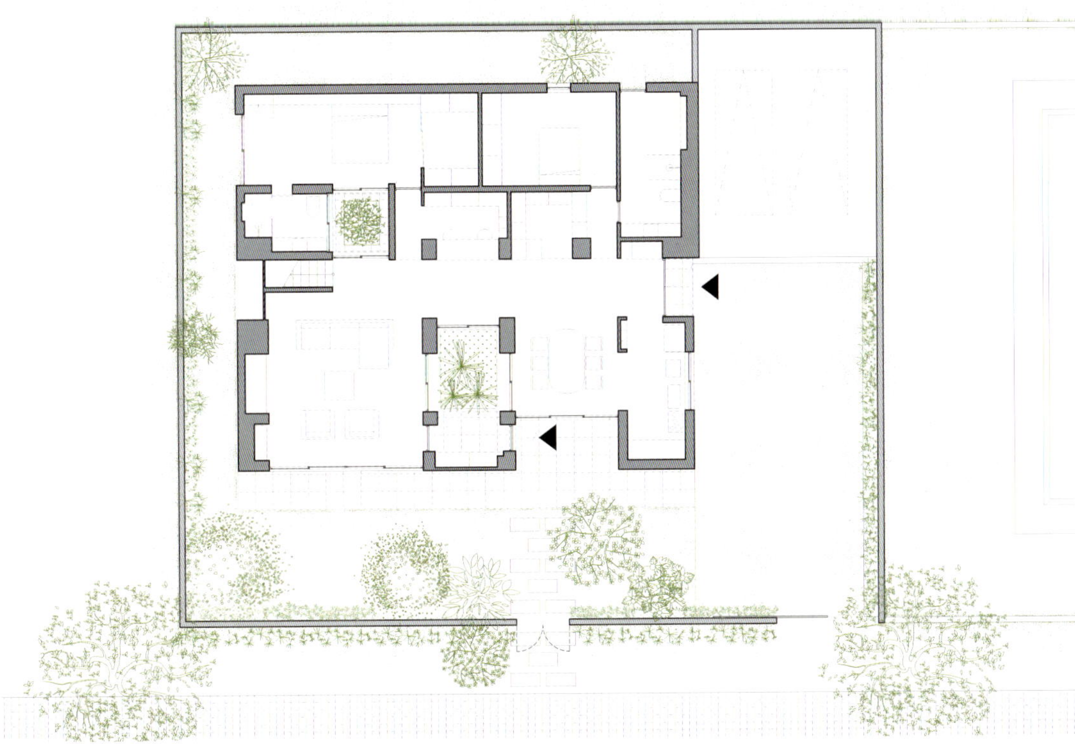

First Floor

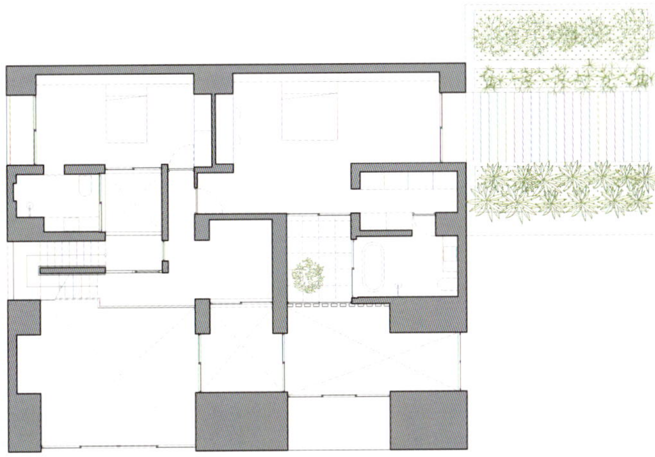

Second Floor

Roof Floor

Bamboo House

In the current Vietnamese context, there are countless narrow alleys with a width of about 2 to 3m, with the same façade architecture style, cold, heavy and closed to ensure the privacy of each family. In addition, with rapid urbanization, big cities in Vietnam are facing the problem of lack of green space and losing connection with nature.

"Bamboo House" was born in that context – this project wants to develop the idea of "tree house," a prototype aimed at providing green spaces in the area. We try to create a comfortable living space in a small and narrow land area, limited in area, only able to get natural light and ventilation through the front and back sides.

The front façade is arranged with random trees, creating an attractive entrance and a cool feeling for the whole house. The bamboo bushes in the façade evoke the familiar image of East Asia, moreover, it also forms a cushion to block out the hot sunlight of South Vietnam. The most important thing is that the green space in this façade creates distance with neighboring houses, creating privacy for the homeowner.

The front and back of the house are open spaces allowing for easy ventilation, saving energy from using air-conditioning for the building.

In addition to creating a green space for the building, bamboo is also used as a formwork for the concrete walls, and the bamboo pattern has created an aesthetic effect and reduces the hardness and heaviness of the traditional concrete wall.

This project provides a solution to create an open living space surrounded by greenery in a very tight and small location. It not only meets functional and aesthetic needs but also connects people with nature.

Sketch

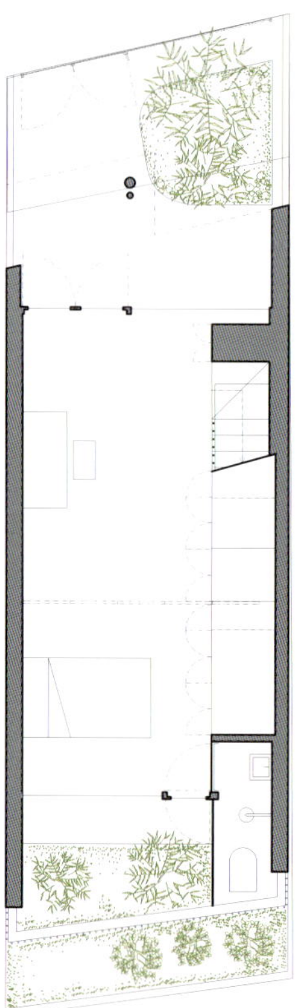

Ground Floor

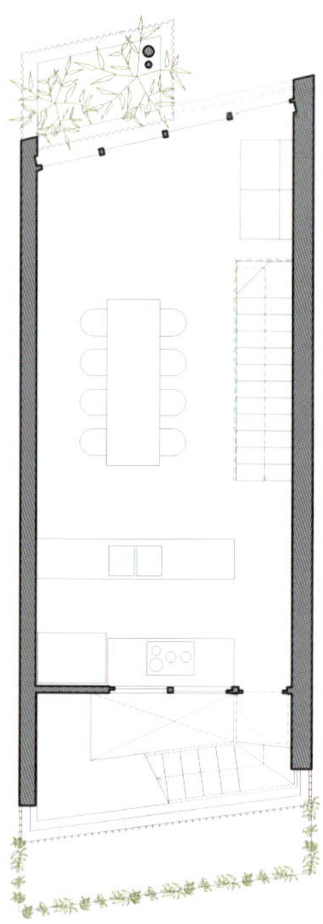

First Floor

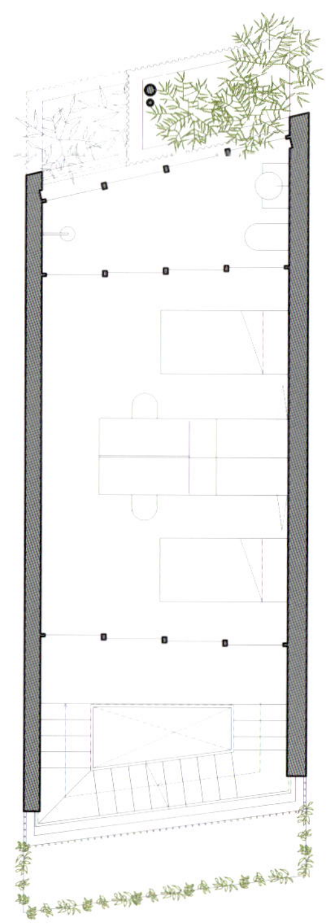

Second Floor

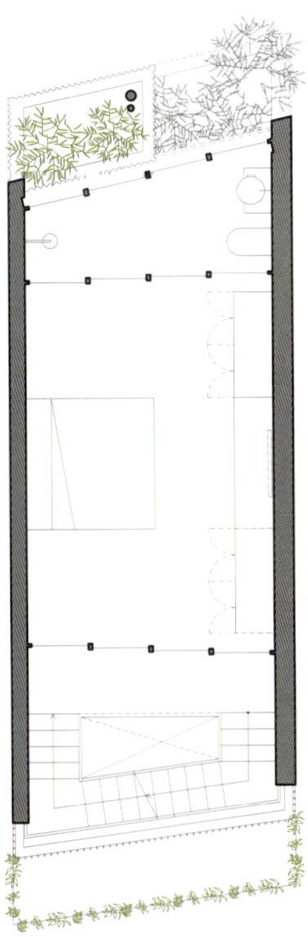

Third Floor

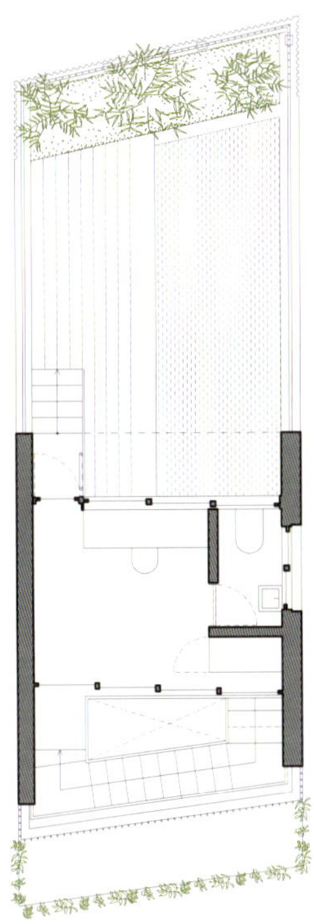

Fourth Floor

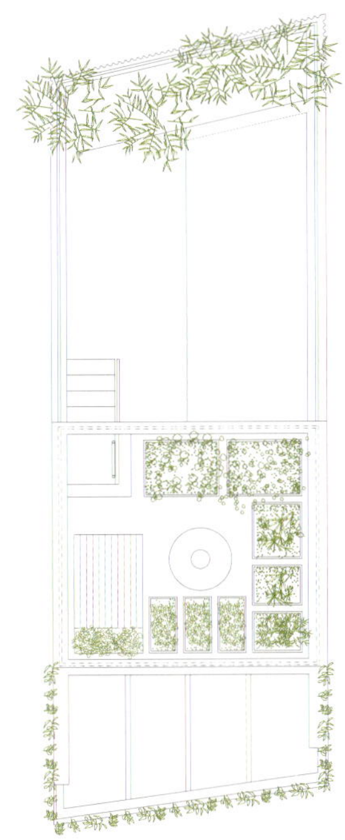

Roof Floor

Binh House

The inhabitant is a family of three generations. The idea was to make a space where people can always interact and communicate, especially encouraging for the elderly.

Gardens are located on the top of every vertical stacking volume, bounded by sliding glass doors. Not only do the shifting stacking openings improve the microclimate by activating natural ventilation and daylight, but they also ensure visibility and encourage interaction between the family members.

Living, dining, bedrooms, and study room are open. From one room, people's sightline can reach beyond to the other room through its garden.

Service areas such as the kitchen, bathrooms, stairs, and corridors are located in the west to limit heat radiation exposure to frequently occupied areas. The vertical variation of spaces creates a lopsided pressure difference. Thus, natural ventilation is ensured even when to-be-built neighboring houses are in place. Thanks to these passive strategies, the house always stays cool in the tropical climate.

The roof gardens host trees for shading, reducing indoor temperatures. The family can grow their own vegetable for daily consumption. This vertical farming solution is suitable for high-density housing and the Vietnamese way of life.

The design reduced operational and maintenance costs thanks to the use of sustainable materials and the microclimate created within the house.

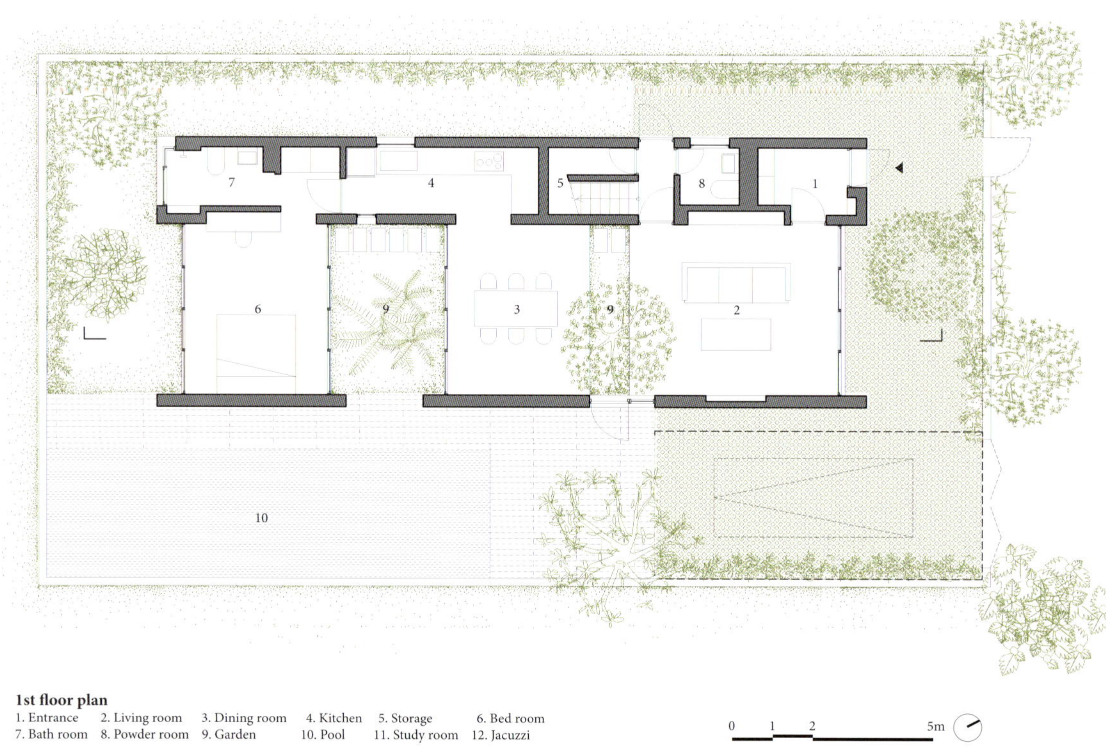

1st floor plan
1. Entrance 2. Living room 3. Dining room 4. Kitchen 5. Storage 6. Bed room
7. Bath room 8. Powder room 9. Garden 10. Pool 11. Study room 12. Jacuzzi

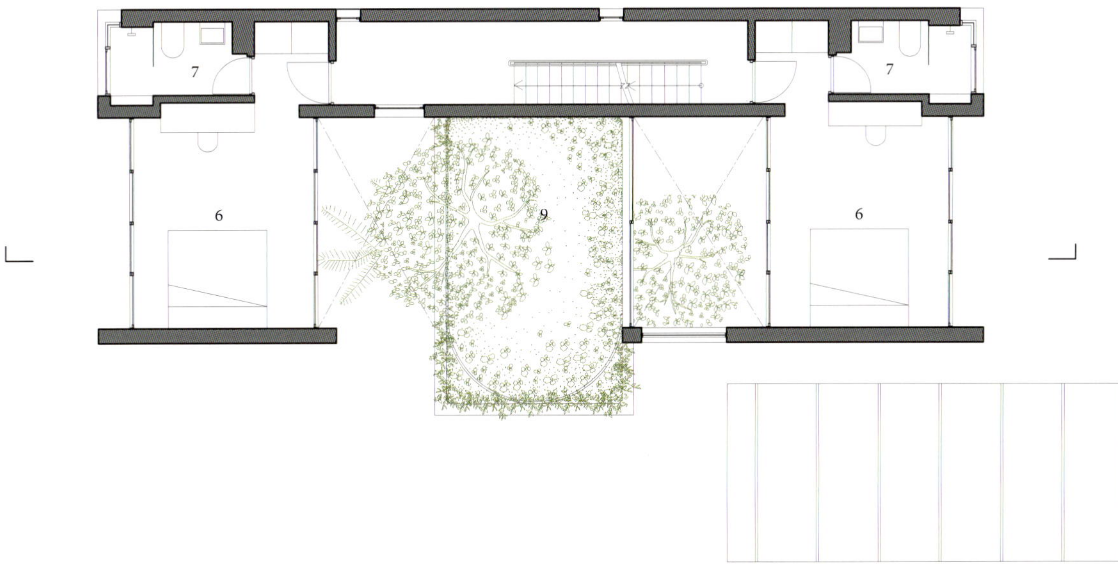

2nd floor plan

1. Entrance 2. Living room 3. Dining room 4. Kitchen 5. Storage 6. Bed room
7. Bath room 8. Powder room 9. Garden 10. Pool 11. Study room 12. Jacuzzi

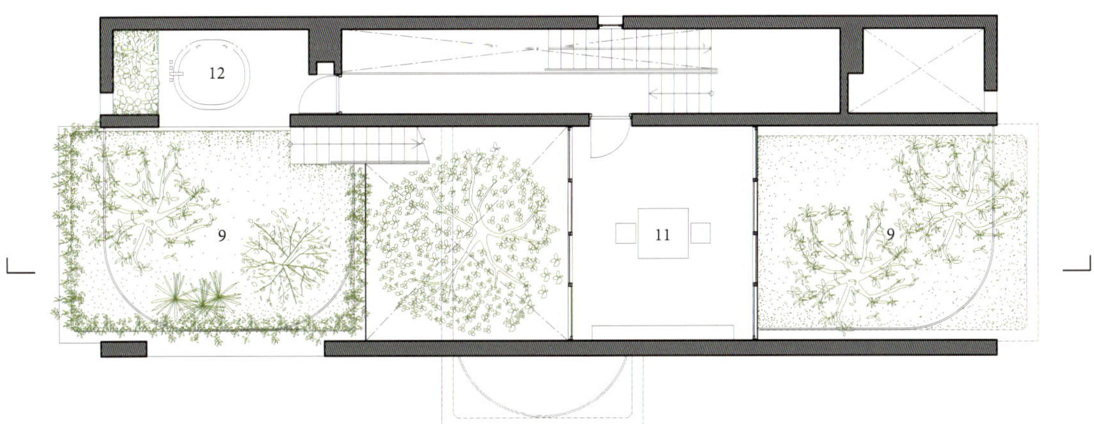

3rd floor plan

1. Entrance 2. Living room 3. Dining room 4. Kitchen 5. Storage 6. Bed room
7. Bath room 8. Powder room 9. Garden 10. Pool 11. Study room 12. Jacuzzi

Roof floor plan

1. Entrance 2. Living room 3. Dining room 4. Kitchen 5. Storage 6. Bed room
7. Bath room 8. Powder room 9. Garden 10. Pool 11. Study room 12. Jacuzzi

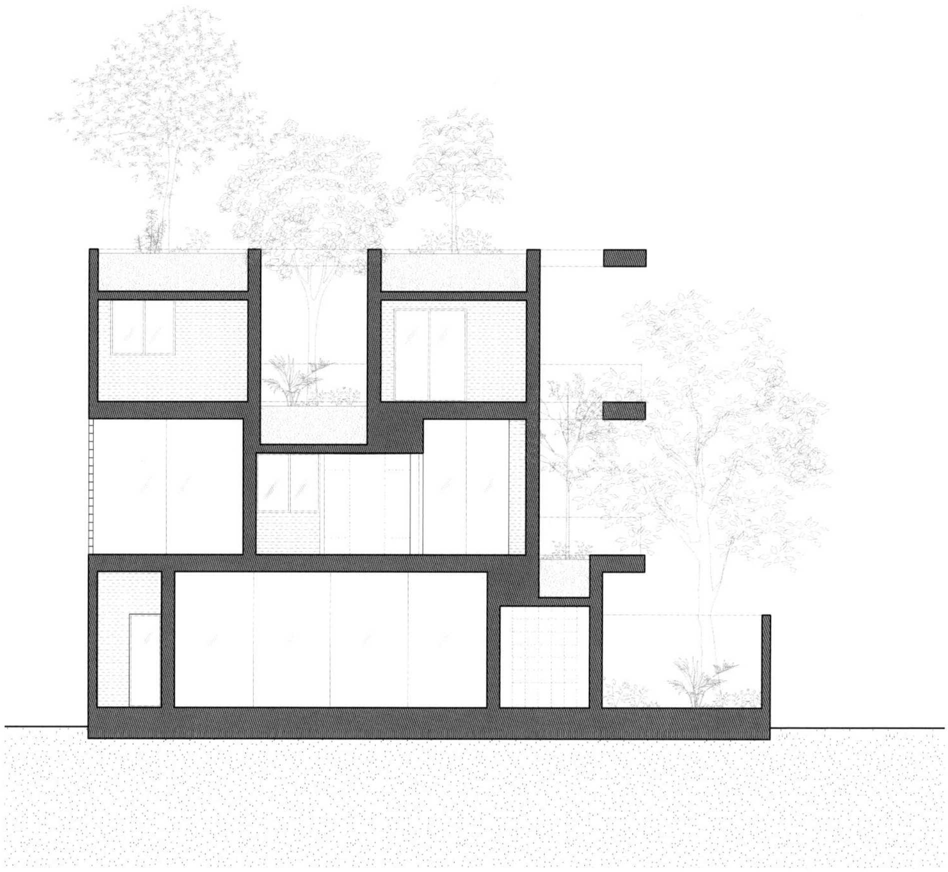

Section

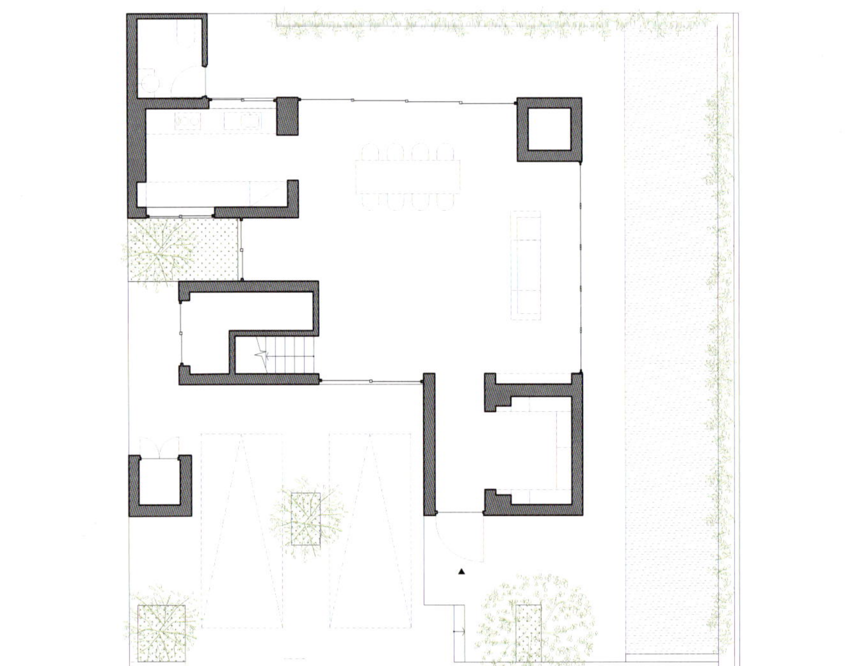

Ground Floor

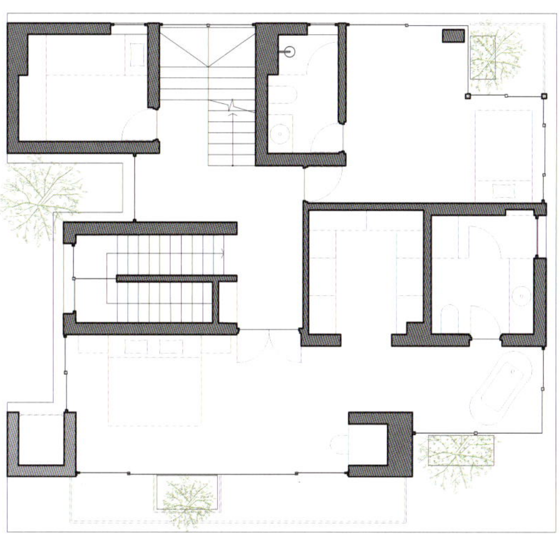

First Floor

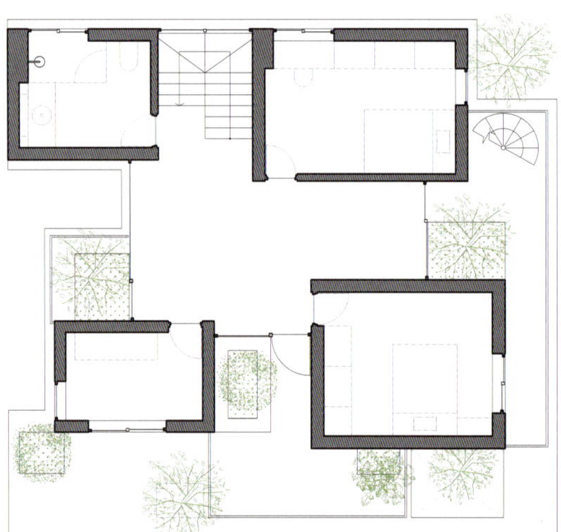

Second Floor

Roof Floor

Stepping Park House

The site is located in a newly developed residential area. On the north side of the house is a green park, which is a unique case not only for VTN Architects but also for a typical Vietnamese resident. The rare opportunity of designing a home next to green public space becomes a realization as we transform the house into an extension of the park.

In section view, Stepping Park House is cut through its three floors by three connected rectangular voids, creating a sense of continuity throughout the spaces, from outside to inside. Each void serves a unique function of the home. The void on the ground floor, in which the living room is situated, opens out to the park. This void connects to the second floor, where the indoor trees are allowed to freely grow upwards. The same designing elements repeat where for the third floor, the void creates an impactful yet elegant visual connection. A skylight that infiltrates the voids is filtered by a series of louvers. Conceptually, the interior of the house becomes an indoor tropical forest.

While the shared living space is as open as possible, the sleeping area of the house remains private as we treat and place them like concrete boxes. However, to provide natural sunlight and ventilation from the outside without jeopardizing privacy, trees are planted in front of the large opening of the boxes, where glass panels are integrated. Not only does the greenery provide shade and reduce direct solar gains, but it also blocks unnecessary viewing from the outside.

As a result of the chimney effect, the connected voids also helps with natural ventilation, thus minimizing the need for air conditioning. Being in the space, one would feel the wind guiding them towards the floors above, leading them through the trees and the ambient skylight filtered by the greenery. Stepping Park House is to become the model for housing in the tropical climate.

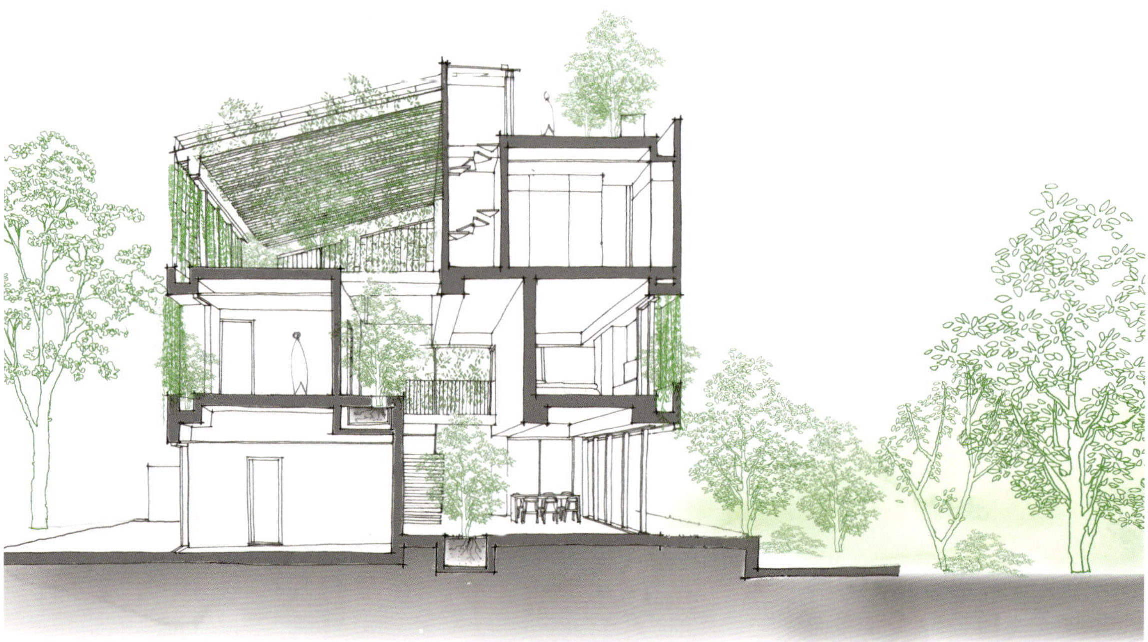

Secion Sketch

Elevation

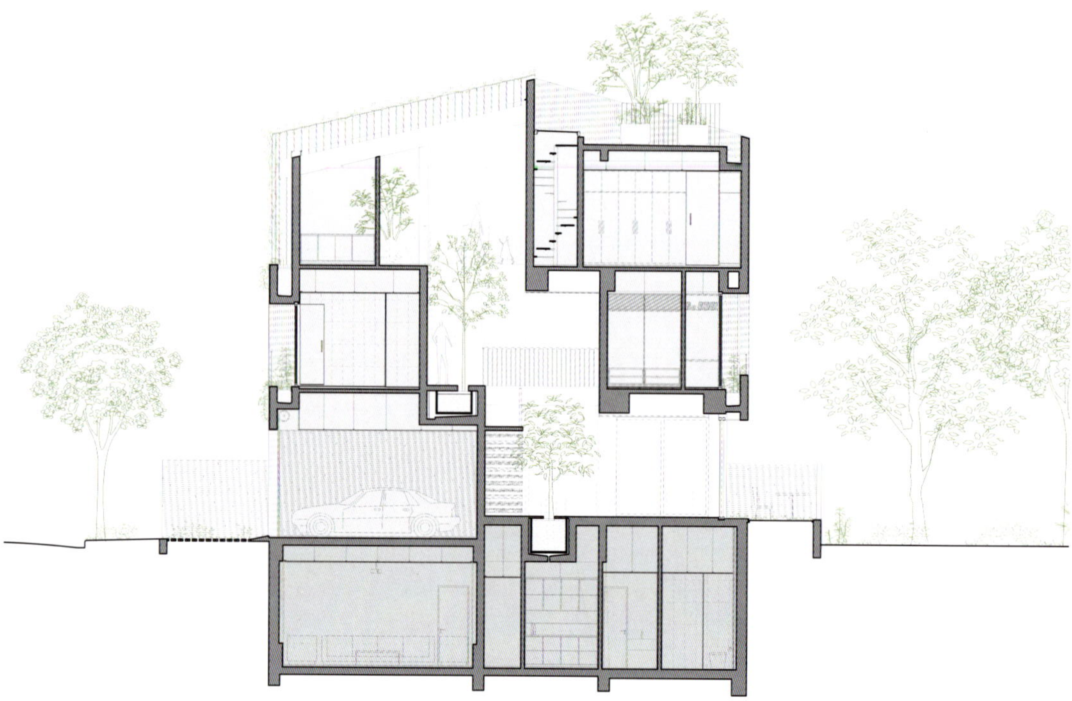

Section

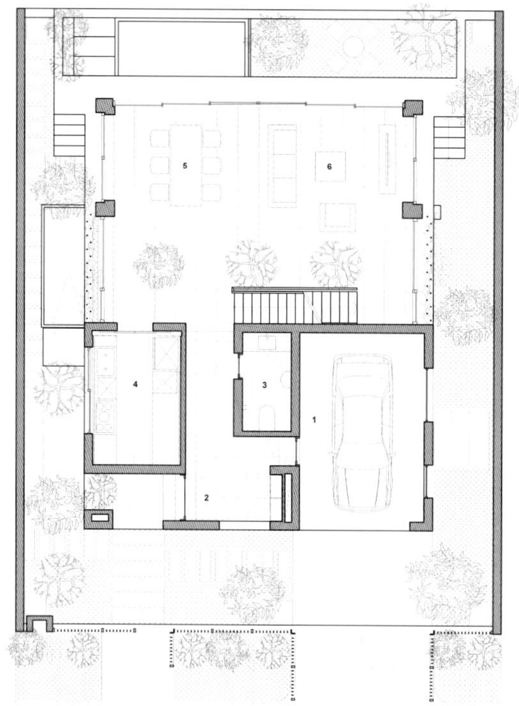

Ground floor plan
1. Garage
2. Entrance Hall
3. WC
4. Kitchen
5. Dining
6. Living room

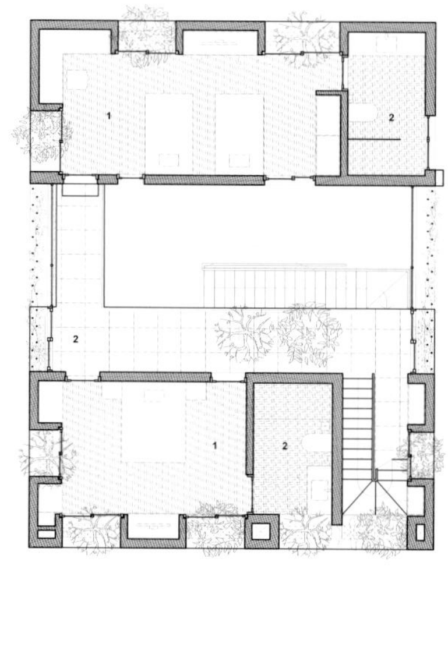

1st floor plan
1. Bedroom
2. Bathroom
3. Circulation

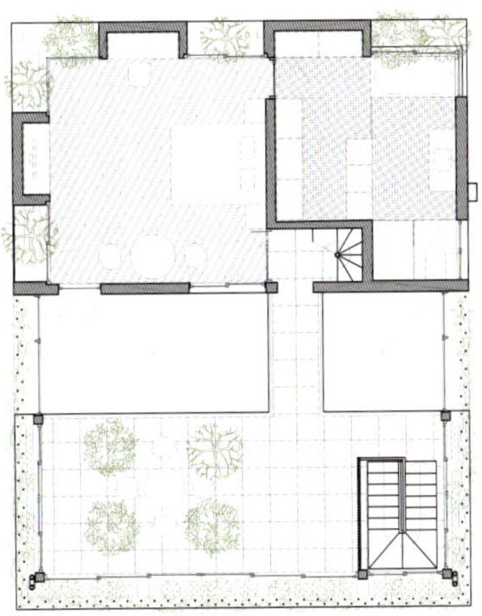

2nd floor plan
1. Master bedroom
2. Bathroom
3. Family room

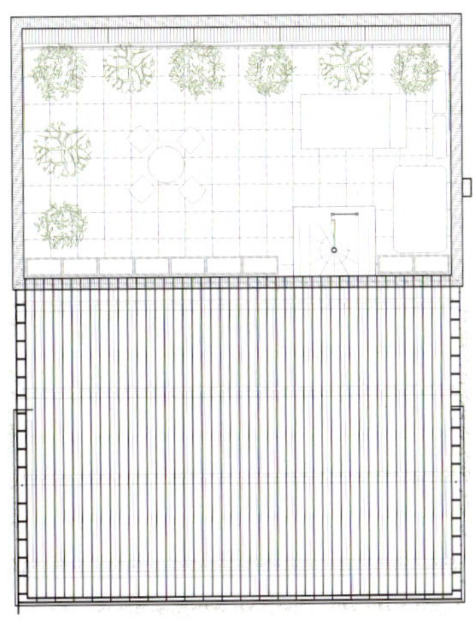

Roof plan
1. Terrace
2. Bathroom
3. Family room

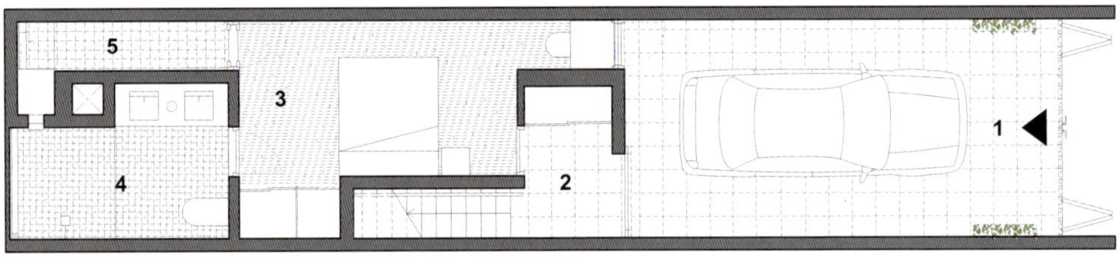

Ground floor plan

1. Garage
2. Entrance Hall
3. Guest room
4. Bathroom
5. Courtyard

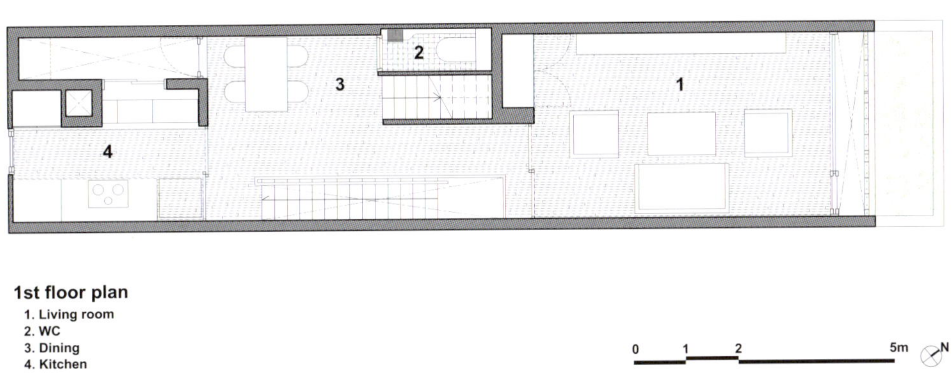

1st floor plan

1. Living room
2. WC
3. Dining
4. Kitchen

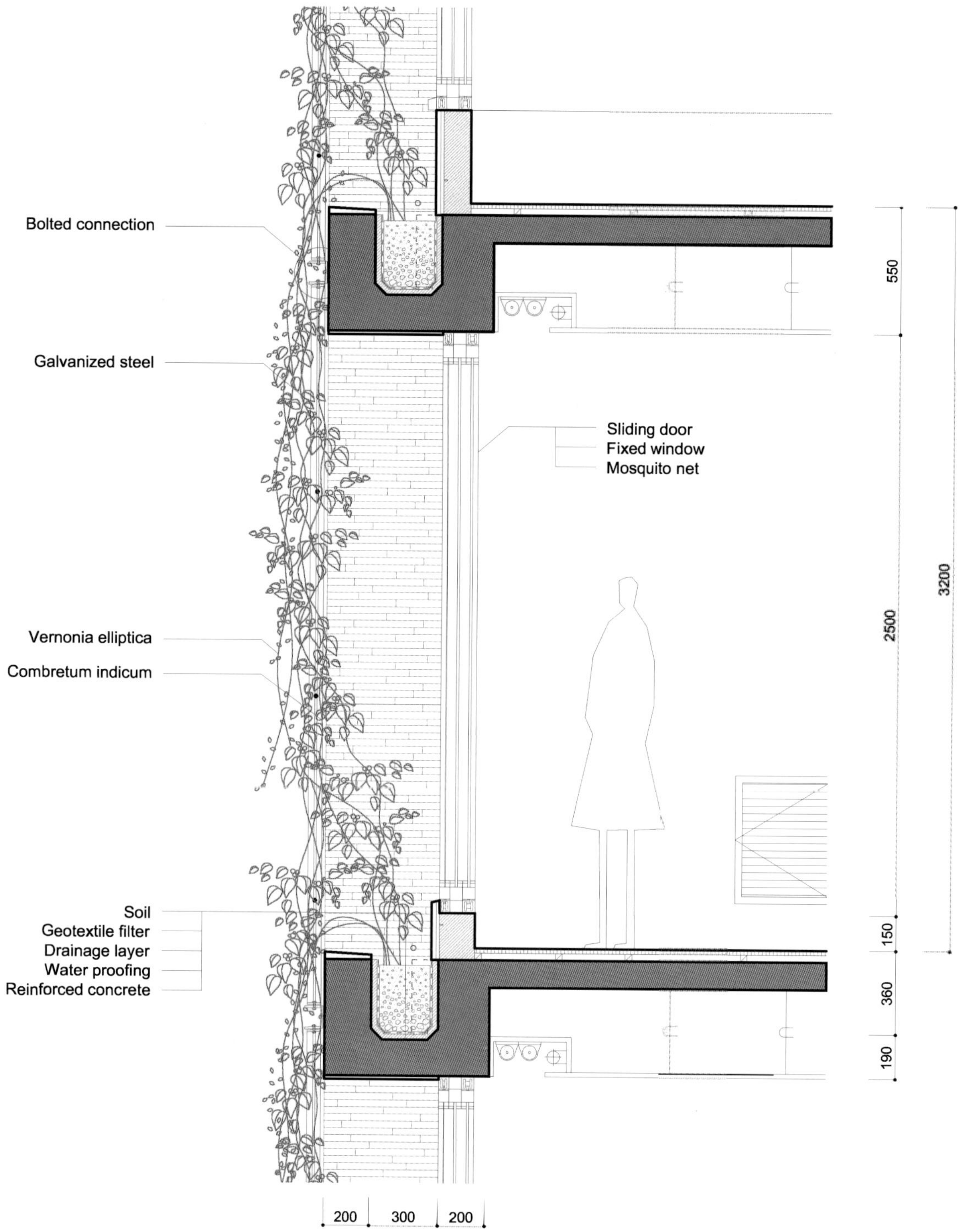

Section Detail

Thang House

Inspired by his childhood imagination of a country house full of rich tropical greenery, grasslands, and fresh air, the owner always had a desire to own a similar house which is located right in the heart of Danang City, Vietnam.

The design team had to propose a solution in limited area of land (250 square meters). The strategy was to create a green lung for the house to breathe by dividing the land into two parallel parts: one for the garden with a huge green wall alongside; other for the living space which has majority of its windows and doors facing towards the green lung with the purpose of bringing natural light, fresh air, and the aroma of grass and flowers to every single corner of the house.

The green area lost due to construction work was compensated by a fruit garden on the rooftop that gets sunlight and shelters the building. The roof garden included 9 tree boxes interspersed with gaps provide more greenery view under natural day and night light. This has not only contributed to the greening of the city but also creates space for urban agriculture to serve daily fresh vegetable and fruit for the family. The perfect combination of farming with a panorama view of an emerging costal city would surely relieve a person from the burden of a hectic city life.

To help the house operate more sustainably and economically, an automatic watering system recycles and circulates water from a fish pond to the roof garden and vice versa, and solar water heating and solar panel system installed on top produces energy for a family sized home. The design aims to make the building self-sufficient in green food production and to minimize energy consumption.

The house comprises four main boxes - one big volume under the roof and three smaller inside the big one. These are extended and hung over the garden to make space for bedrooms and living room. While the big box is covered by grey stone for a deep and charming feeling of the common space, the small ones are covered by white brick for an enlightening and warm feeling of living space. People can travel through different spatial in-house atmospheres. The narrow roof gaps allow the entry of natural light subtly and sufficiently for the interior space. It creates an effect playing with light through the void between three levels.

This house is mainly made of local building materials which are available from nearby sources, such as the grey stone from Hoa Son rockery mine and brick from Quang Nam Province. In addition, the exposed concrete is an efficient material for low-tech construction in Vietnam and is durable under the harsh local climate.

Under rapid urbanization, the lack of greenery in Vietnam cities generates various urban problems: increasing air pollution without purification, the city's inability to retain rainwater, insufficient amount of space to cool the street, and people become inactive over time. Thang House is one of the buildings under the "House for Trees" series and it pledges to contribute to the greening of the urban landscape and improving the habitat by providing sustainable and environmentally friendly architecture. Thang House is yearly covered by tropical fruit trees which creates a tranquil, peaceful space for every family member and simultaneously preserves a green corner in the middle of this rapidly developing city.

Section Sketch

Section Sketch

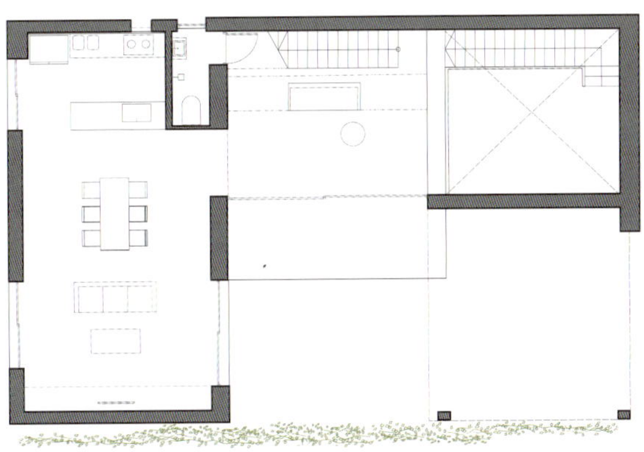

Second Floor

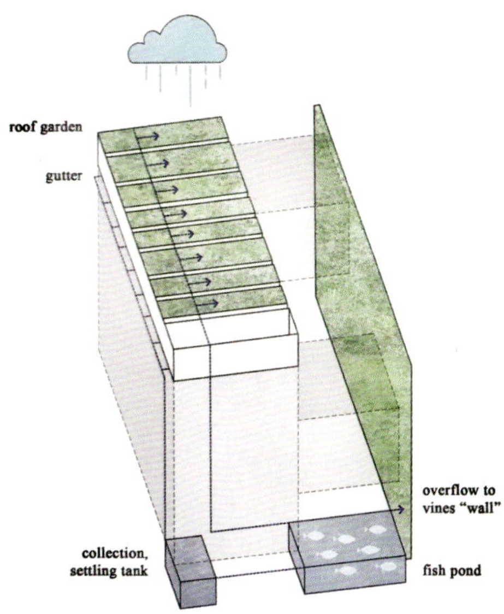

Water recycle diagram

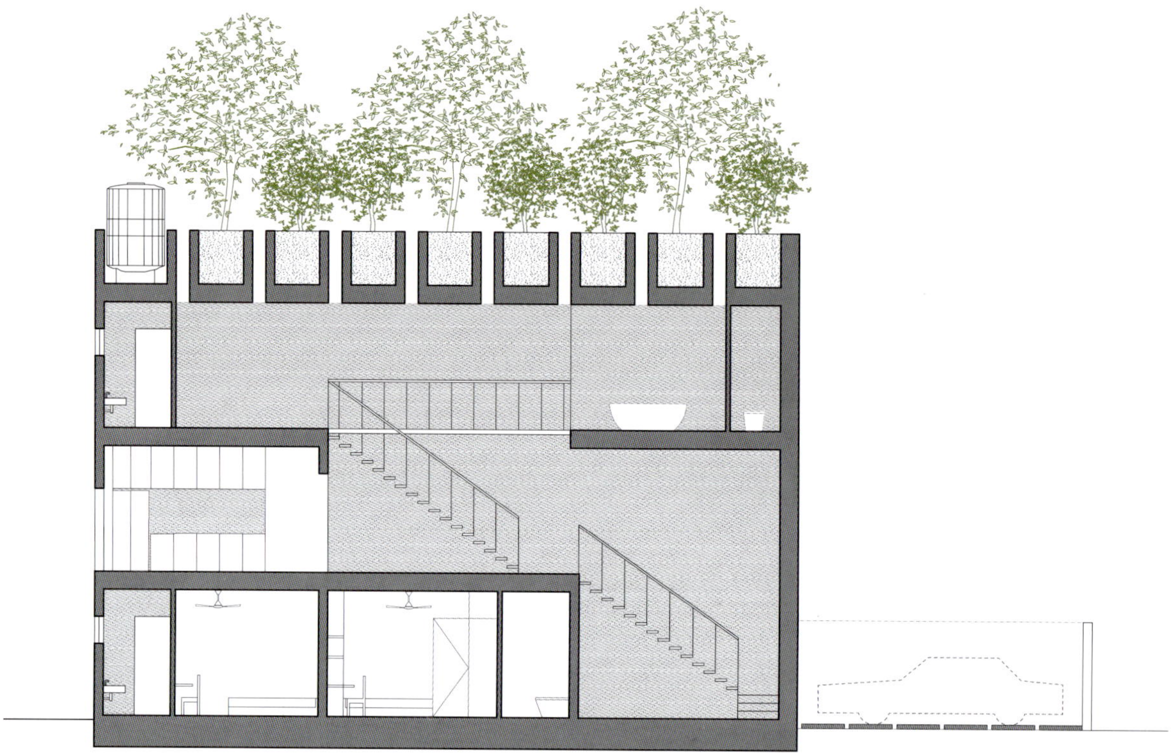

Section

Site Plan

Elevation

F20

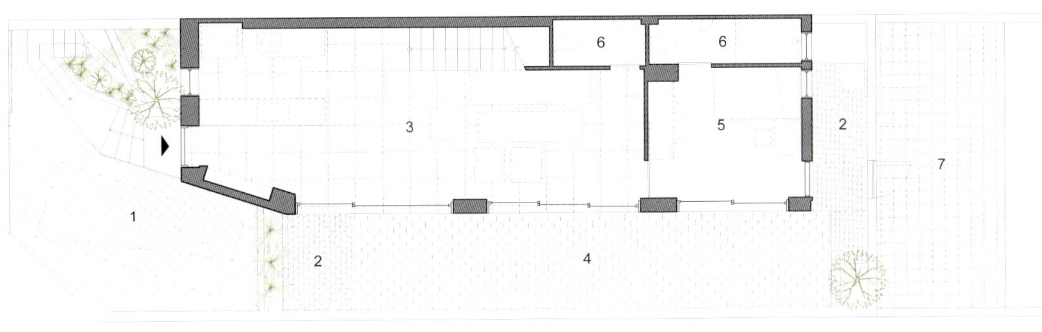

Ground Floor

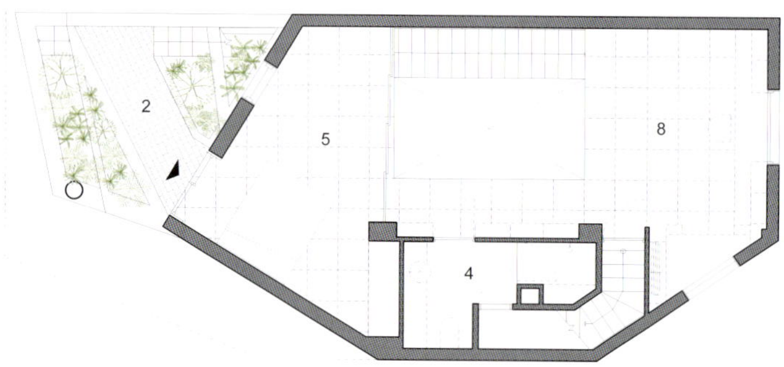

First Floor

ABC
ULTIMATE BUGOPEDIA

Axonometric Sketch

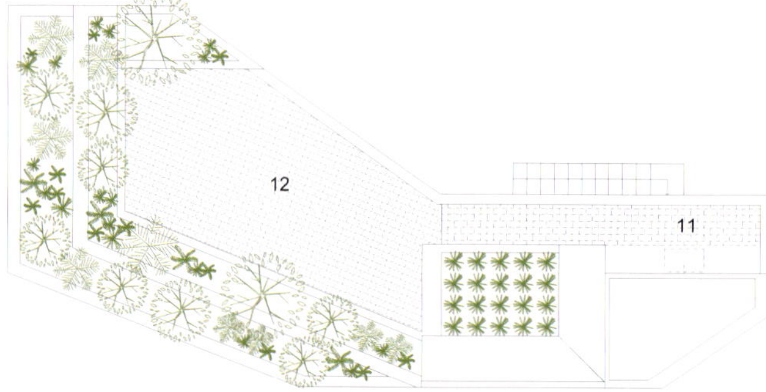

Roof Floor

Ha Long Villa

Vietnam traditionally owns numerous natural and cultural areas that are registered as World Heritage Sites, among them the well-celebrated Ha Long Bay. Economic growth driven by tourism has raised the local standard of living but also given away the land's natural characteristics to reckless development by phasing out into the water surfaces. Ha Long, home to Ha Long Villa, is a beautiful coastal city in northern Vietnam, overlooking Ha Long Bay's breathtaking landscape.

In such a rich setting, we wanted to designed a sustainable home that exists in harmony with the wonderful surrounding environment and seeks to become part of its landscape. The main concept of the house was to create a space where people can feel what it is like to live in a forest. (Hence, the main concept of the house was to replicate one of Ha Long Bay's towering limestone islands topped by rainforest.)

In order to create a forest-like atmosphere and recreate the island atmosphere, the house was formed from a pentagon within another pentagon that establishes spatial layers of interior and semi-exterior spaces that were filled with greenery. This composition creates deep shadows cast by the double skin green façade. The buffer space between the interior and exterior spaces protects the villa from the area's harsh climate and noise. Every semi-exterior part of the home is connected by the main spiral staircase, creating a spatial sequence from the outside to the inside, from ground to roof. Large windows overlooking the semi-outdoor extents provide views to the dramatic bay and cityscape from different perspectives.

The semi-exterior space connects the interior and the terrace, and is a place to promote various living activities. Not only does the multi-purpose space serve as a circulation to all the rooms, it is a place for gardening, sitting, or just simply wandering around. These distinctive spaces offer residents options in their daily lives, to dine inside or outside on a particular day for example. The daily activities of the residents can be observed through the generous openings.

Exterior walls are made by rough exposed concrete, creating an impression of the lime stones found in Ha Long Bay. The greenery in the semi-exterior spaces casts shadows on this strong facade, creating an ever-changing appearance over time. The contrast between the rough concrete and soft trees expresses a unique impression. Additionally, with plants on the roof, we create an area several times larger than the original green area.

As part of the vast landscape, Ha Long Villa harmonizes with the spectacular heritage site of Ha Long Bay. And Ha Long Villa aims to be a space where people return to the tradition of living with nature.

Bim One

Section 2

1. Main entrance 2. Sub - entrance 3. Garden 4. Garage 5. Laundry room 6. Toilet 7. Entertainment room 8. Closet 9. Maid room 10. Terrace 11. Living room 12. Dining room 13. Kitchen 14. Bedroom 15. Bathroom 16. Patio 17. Altar room 18. BBQ 19. Technical room 20. Vegetable garden

5th floor plan

1. Main entrance 2. Sub - entrance 3. Garden 4. Garage 5. Laundry room 6. Toilet 7. Entertainment room 8. Closet 9. Maid room 10. Terrace 11. Living room 12. Dining room 13. Kitchen 14. Bedroom 15. Bathroom 16. Patio 17. Altar room 18. BBQ 19. Technical room 20. Vegetable garden

0 1 2 5m

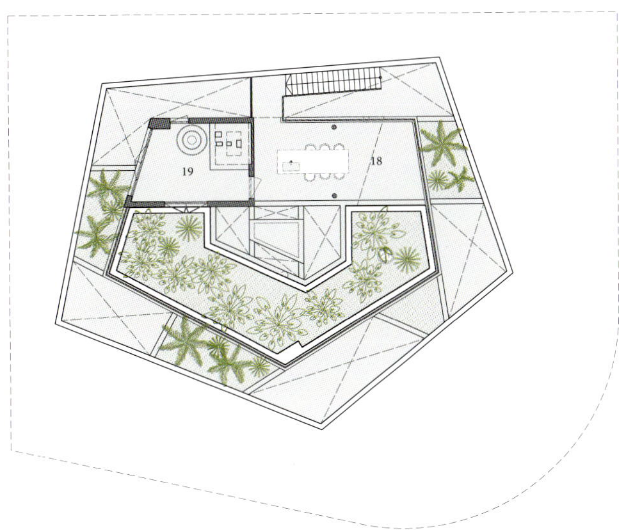

Roof floor plan

1. Main entrance 2. Sub - entrance 3. Garden 4. Garage 5. Laundry room 6. Toilet 7. Entertainment room 8. Closet 9. Maid room 10. Terrace 11. Living room 12. Dining room 13. Kitchen 14. Bedroom 15. Bathroom 16. Patio 17. Altar room 18. BBQ 19. Technical room 20. Vegetable garden

0 1 2 5m

CƠ SỞ SẢN XUẤT GỐM SỨ MỸ NGHỆ CỔ TRUYỀN BÁT TRÀNG
Đạo Xuân
YÊN LỌ LỘC BÌNH - NẬM RƯỢU
Nhận hợp đồng theo yêu cầu quý khách!
HUY
HỒNG
TIẾN
NGA
THỊT

Bat Trang House

The project is situated in a unique location of Bat Trang Town – a pottery village that has been around for more than 10 decades. The façade of the building represents the community's unique culture of ceramic pottery making. While the exterior is inspired by the traditional beauty of the town, the interior ensures the convenience and comfort of today's modern life. The spatial arrangement of the house is also based on the owner's ideal home, where nature intertwines with other functions of the house.

The brief of the project is to design a home for a seven-member family and a place for relative get-togethers. Besides its residential uses, Bat Trang House also functions as a shop, trading the traditional ceramic products of the pottery town.

In the cultural-rich context of the village, using local materials was prioritized by the design team. Additionally, the owner used to be a ceramic specialist, hence VTN Architects is able to take advantage of the material. Ceramic has its own rough aesthetic and high durability, which was suitable for the exterior skin of the building. The entire façade of the design is made purely out of ceramic bricks. The bricks are individually made to size to create the alternating rhythm, opening up in some spaces while closing in others. The small gaps created by the bricks' placement act as vents, circulating the air throughout the building. The big openings of the façade situate sizable greeneries, allowing filtered light to flow through.

The interior spaces are designed with a minimalistic approach. Large glass panels are integrated without the fear of getting the house overheated, as direct sunlight is filtered by the two layers of ceramic façade and greenery façade. The gaps of the ceramic walls function particularly well in terms of effectively getting the right amount of solar gain for the interior, especially in the summer. Artificial air conditioning is unnecessary as the three-layer ventilation system "exterior ceramic façade," "alternating green spaces," and "door" guarantee that the house is "absolutely cool." Solar panels and rainwater collectors also take part in the sustainable design. Vast windows, and alternating ceramic openings connected by the indoor forest create a multi-layer living picture frame that would be ever-changing for the occupants.

The intimate connection of people with nature is notably important to the design team. VTN Architects ensures the natural and sustainable living quality in every residential project. The designs make sure that the occupants could always feel the freshness of greeneries and the ambient skylight in their daily activities.

Elevations

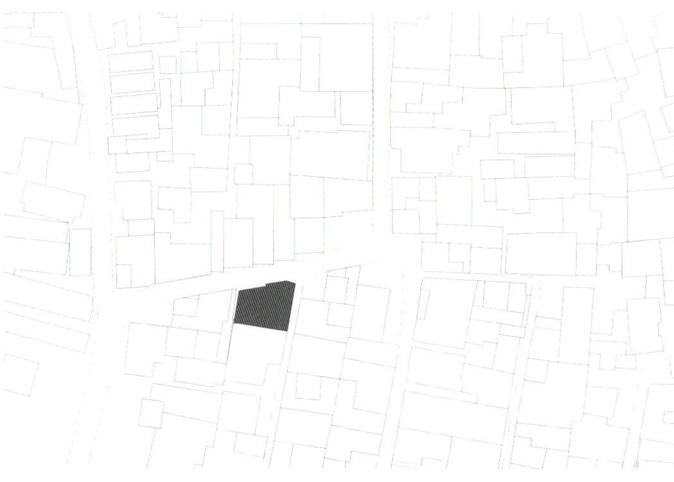

Site Plan

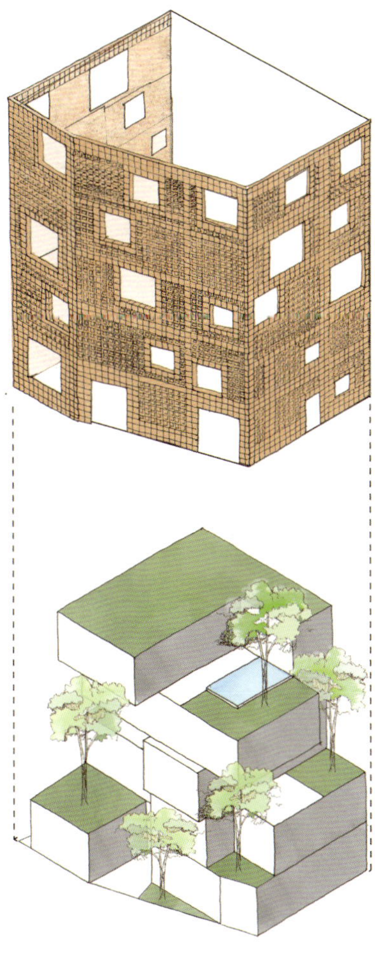

Diagram

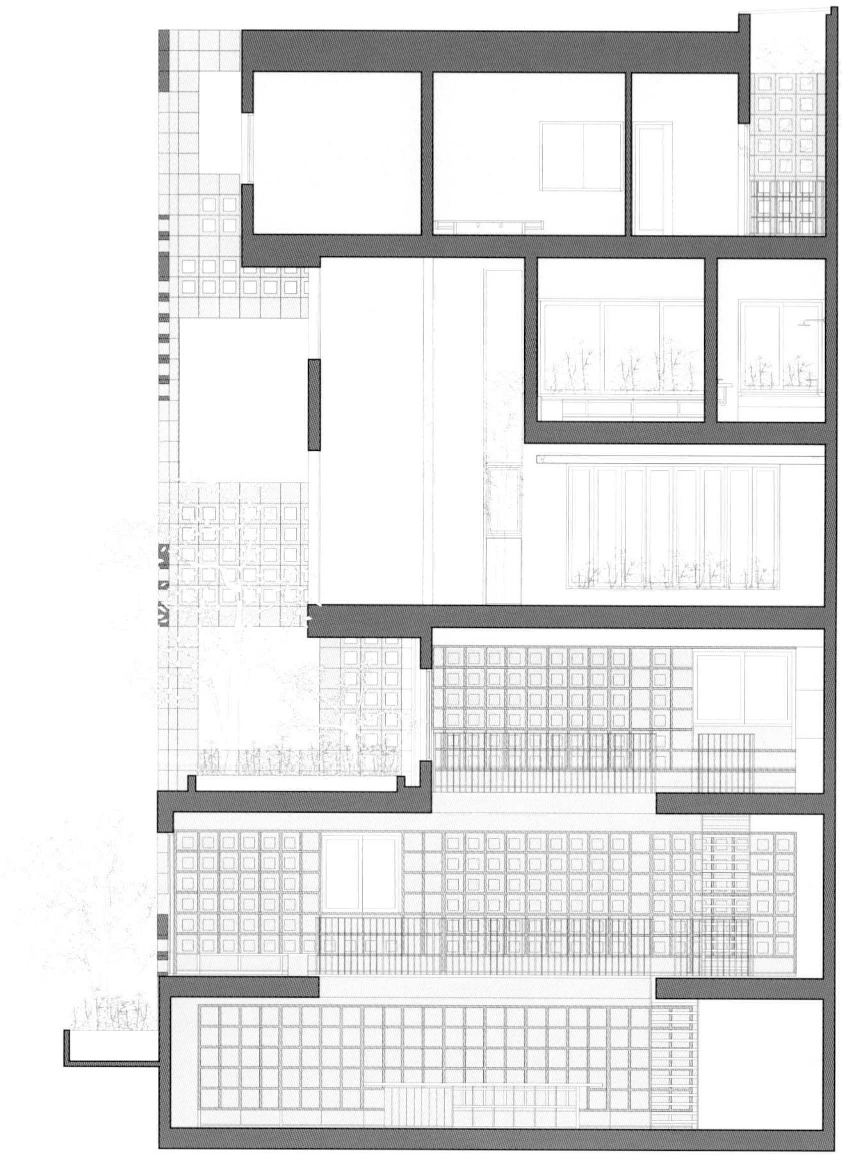

Section

S House 1-2-3

S House 1

This serial project called S House aims to provide stable, lightweight, permanent, and affordable houses for low-income residents in harsh tropical climates. By minimizing the functions of the house and using low cost materials, the construction cost of a house can be reduced to as little as 3200USD. Passive design methods are applied throughout the series, including but not limited to natural lighting and ventilation.

Two prototypes were erected in Dong Nai province, on the construction site of a kindergarten. The first prototype, with a floor area of 22.5 square meters, was designed as a model home. The second, at roughly 18 square meters, was designed as an office for the kindergarten, showing the flexibility of this prototype.

On the assumption that the bathroom and kitchen are placed outside and shared with several families, the house has minimal space for living, eating, and sleeping. The plan was designed to be adjustable towards the longitudinal axis, allowing for future expansion. Its interior is a simple one-room space, articulated by curtains and fluctuations in levels. The floor rises higher in one part to become furniture.

Another distinctive feature is the use of folding beds, which allows the dwellers to sit on the floor during the day. These beds can also be transformed into sofas. In order to reduce the construction cost, dwellers were encouraged to participate in the construction process. The structure of the prototype is a lightweight steel frame, which is easy to assemble without the need of heavy machines nor special techniques.

Considering the recyclability of materials, wet joints are avoided. The roof is supported by truss-beams of steel bars, which minimize steel material and provide an ideal pitch for water drainage.

The envelope of the house is composed of polycarbonate wall panels and corrugated FRP roof panels, with bamboo louvers set inside. Both materials are available everywhere in Vietnam and they are cheap, light, and replaceable. Translucent envelope and bamboo louvers filter the harsh direct sun in tropical climates. The interior is filled with diffused light and reduces the need for artificial lighting, which in turns drastically reduces electricity consumption. There is also a gap between the roof and the wall, which evacuates hot air. As the whole space is naturally ventilated, there is no need for air conditioning. A pent roof was designed to collect rain water for daily use in the dry season. Utilizing nature, dwellers can save a great amount of energy, resources and money.

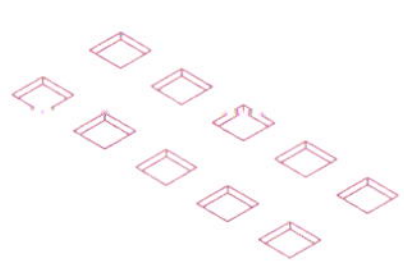

1. Digging

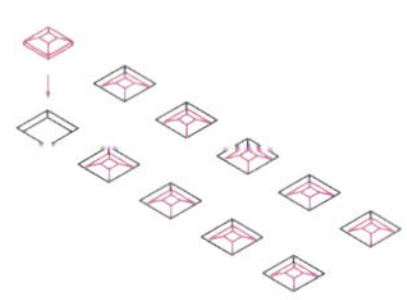

2. Placing Foundation

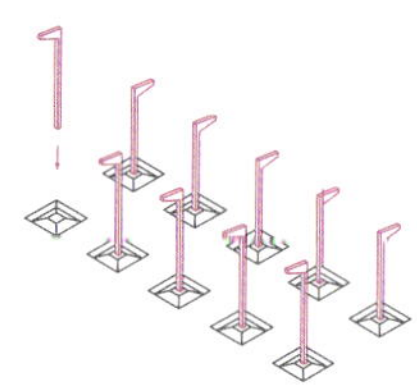

3. Errecting Column

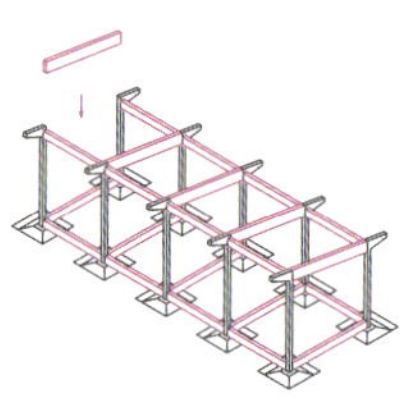

4. Beam

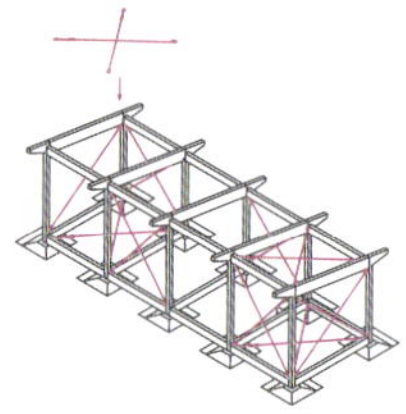

5. Steel Braces

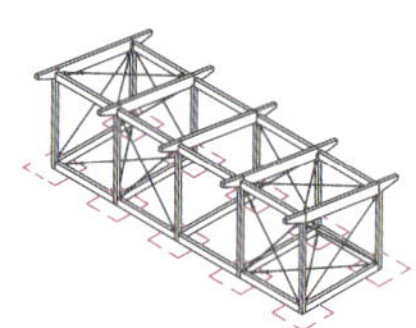

6. Covering Footing

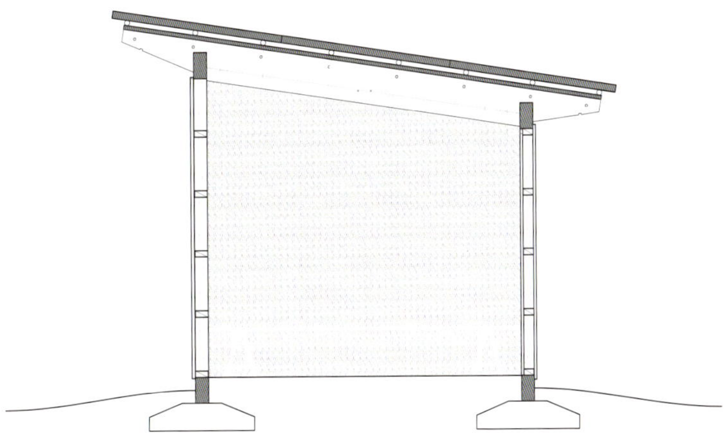

Section

S House 3

After building two successful case-study houses, a steel structure model in 2012 and a precast concrete structure model in April 2014, a new prototype, S House 3, was constructed in Ho Chi Minh City.

Combined with the best features from the first and second prototype, S House 3 realized the highest durability, and flexibility with the easiest construction while maintaining affordability and lightness. Slender steel lattice walls secure, support, and rigidify the whole frames, reducing the number of beams needed. Hence, the structure only weighs roughly 1,200 kg. A pedestal, incorporated with the foundation, is made of reinforced concrete. As a result, the galvanized steel structure was isolated from the ground, ensuring its durability even in a high-temperature, high-humidity climate.

Similar to the previous prototypes, the basic strategy of construction was a combination of modular structural components and DIY finishes. However, all components were designed to be lighter than 60 kg, allowing easier transportation and construction. It could take only three hours to assemble the whole structure and roof.

Finishing materials can all be assembled and replaced easily by dwellers and neighbors. For the prototype, a lightweight cement board was chosen for cladding.

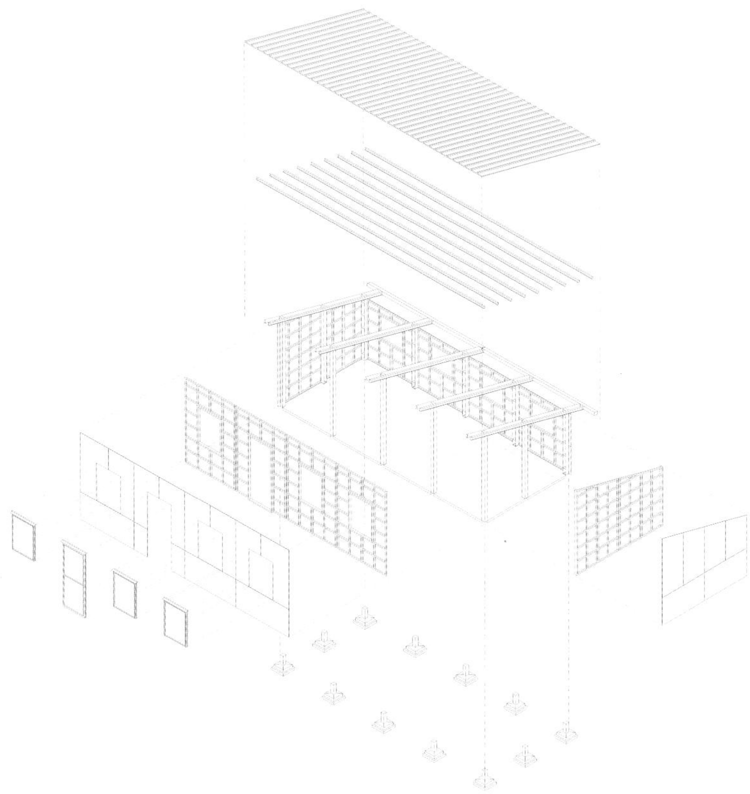

Exploded Axonometric

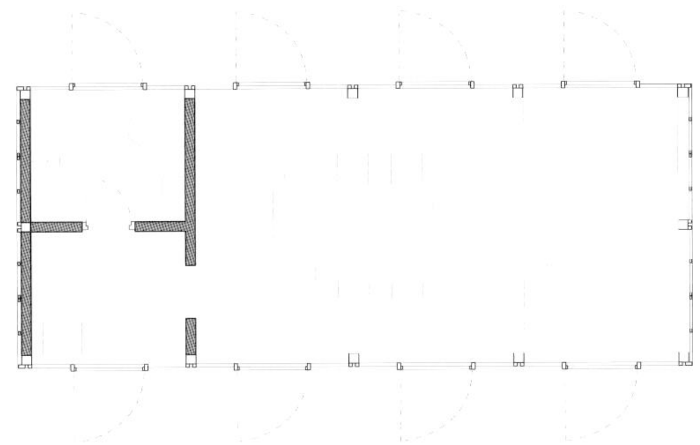

Ground Floor

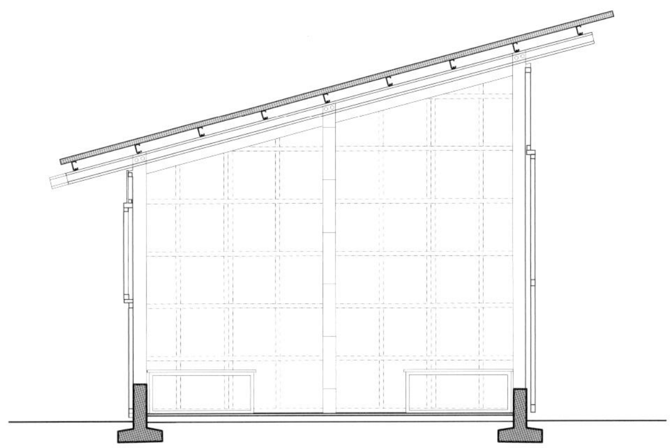

Section

Projects Credits

Stacking Green

Principal Architects	Vo Trong Nghia, Daisuke Sanuki, Shunri Nishizawa
Design Team	Nguyen Minh Tung
Contractor	Wind and Water House JSC
Status	Built in 02.2011
Program	Private house
Location	District 2, Ho Chi Minh City
GFA	215 sqm
Site Area	1,160 sqm
Photographs	Hiroyuki Oki
Client	Individual

Hoan House

Principal Architects	Vo Trong Nghia, Masaaki Iwamoto
Status	Built in 03.2015
Program	Private house
Location	Nha Trang, Vietnam
GFA	331.9 sqm
Site Area	492.0 sqm
Photographs	Hiroyuki Oki
Client	Individual

Gia Lai House

Principal Architects	Vo Trong Nghia, Daisuke Sanuki, Shunri Nishizawa
Design Team	Phan Chau Trinh
Contractor	Wind and Water House JSC
Status	Built in 10.2011
Program	Private house
Location	Gia Lai, Vietnam
GFA	510 sqm
Photographs	Hiroyuki Oki
Client	Individual

Bamboo House

Principal Architects	Vo Trong Nghia
Architect	Kuniko Onishi
Status	Built in 2016
Program	Private house
Location	Ho Chi Minh, Vietnam
GFA	217 sqm
Photographs	Hiroyuki Oki

Binh Thanh House

Principal Architects	Vo Trong Nghia, Shunri Nishizawa, Daisuke Sanuki
Design Team	Le Thi Anh Huyen, Nguyen Tuan Anh
Contractor	Wind and Water House JSC + Danang Company
Status	Built in 06.2013
Program	Private house
Location	Binh Thanh, Ho Chi Minh City, Vietnam
GFA	516 sqm
Site Area	140 sqm
Photographs	Hiroyuki Oki
Client	Individual

Binh House

Principal Architect	Vo Trong Nghia
Design Team	Masaaki Iwamoto, Chiang Hsing-O, Nguyen Tat Dat, Nguyen Duy Phuoc, Takahito Yamada
Contractor	Wind And Water House JSC
Status	Built in 03.2016
Program	Private house
Location	District 2, Ho Chi Minh City
Site Area	322 sqm
Footprint	136 sqm
GFA	233 sqm
Photographs	Hiroyuki Oki, Quang Dam

House for Trees

Principal Architects	Vo Trong Nghia, Masaaki Iwamoto, Kosuke Nishijima
Design Team	Nguyen Tat Dat
Contractor	Wind and Water House JSC
Status	Built in 04.2014
Program	Private house
Location	Tan Binh District Ho Chi Minh City
Site Area	475 sqm
Footprint	112 sqm
GFA	226 sqm
Photographs	Hiroyuki Oki
Client	Individual

Stacked Planters House

Principal Architect	Vo Trong Nghia
Design Team	Masaaki Iwamoto, Nguyen Quynh Han, Kuniko Onishi
Contractor	Wind and Water House JSC
Status	Built in 2017
Program	Private house
Location	Ho Chi Minh, Vietnam
Site area	233.6 sqm
GFA	260.8 sqm
Photographs	Hiroyuki Oki
Client	Individual

Stepping Park House

Principal Architects	Vo Trong Nghia, Hidetoshi Sawa
Design Team	Nguyen Van Thien
Contractor	Wind and Water House JSC
Status	Built in 2018
Program	Private house
Location	Ho Chi Minh City, Vietnam
Site Area	252 sqm
GFA	475 sqm
Photographs	Hiroyuki Oki
Client	Individual

Breathing House

Principal Architects	Vo Trong Nghia, Kosuke Nishijima
Design Team	So Adachi
Contractor	Wind and Water House JSC
Status	Built in 2019
Program	Private house
Location	Ho Chi Minh City, Vietnam
Site area	69.5 sqm
GFA	343 sqm
Photographs	Hiroyuki Oki
Client	Individual

Thang House

Principal Architect	Vo Trong Nghia
Design Team	Le Phuong Uyen, Kosuke Nishijima
Contractor	Nguyen Van Thang
Status	Built in 2019
Program	Private house
Location	Da Nang, Vietnam
GFA	250 sqm
Photographs	Hiroyuki Oki
Client	Individual

Ha House

Principal Architects	Vo Trong Nghia, Tran Thi Hang
Design Team	Takahito Yamada, Le Viet Minh Quoc, Mitsuyoshi Shingu
Contractor	Wind and Water House JSC
Status	Built in 2019
Program	Private house
Location	Ho Chi Minh, Vietnam
Site area	137.2 sqm
Footprint	100.5 sqm
GFA	174.8 sqm
Photographs	Hiroyuki Oki
Client	Individual

Ha Long Villa

Principal Architect	Vo Trong Nghia
Design Team	Nguyen Van Thu
Status	Built in 2020
Program	Private house
Location	Ha Long City, Quang Ninh Province
Site area	514.0 sqm
Footprint	231.5 sqm
GFA	1190.0 sqm
Photographs	Hiroyuki Oki

Bat Trang House

Principal Architect	Vo Trong Nghia
Design Team	Ngo Thuy Duong, Nguyen Van An, Do Huu Tam, Pham Phuong Thao
Status	Built in 10.2020
Program	Private house
Location	Bat Trang, Hanoi, Vietnam
Site area	220 sqm
GFA	740 sqm
Photographs	Hiroyuki Oki
Client	Individual

S House 1, 2, 3

S House 1	
Principal Architects	Vo Trong Nghia, Masaaki Iwamoto
Architect	Kosuke Nishijima
Contractor	Wind and Water House JSC
Status	Built in 08.2012
Program	Prototype house
Location	Dong Nai, Vietnam
GFA	House no.1: 22.5 sqm, House no.2: 18 sqm
Photographs	Hiroyuki Oki
Client	Wind and Water House JSC
S House 2	
Principal Architects	Vo Trong Nghia, Masaaki Iwamoto, Kosuke Nishijima
Contractor	Wind and Water House JSC
Status	Built in 04.2014
Program	Prototype house
Location	Long An, Vietnam
GFA	33.1 sqm
Site area	1295 sqm
Photographs	Hiroyuki Oki
Client	Wind and Water House JSC
S House 3	
Principal Architects	Vo Trong Nghia, Masaaki Iwamoto, Kosuke Nishijima
Architect	So Adachi
Contractor	Wind and Water House JSC
Status	Built in 11.2014
Program	Prototype house
Location	Ho Chi Minh City, Vietnam
GFA	31.6 sqm
Photographs	Hiroyuki Oki

Biography

Vo Trong Nghia

Biography Vo Trong Nghia was born in 1976 in Quang Binh province, Central Vietnam. He moved to Japan in 1996 as a Japanese government's scholarship student and started studying architecture. After graduation from Nagoya Institute of Technology in 2002, he joined the University of Tokyo's Landscape and Civic Design Laboratory under the Department of Civil Engineering. There he studied under Prof. Hiroshi Naito. He received his master's degree in 2004. His research thesis discussed the utilization of wind in Hoi An, Vietnam and was awarded the president's award of Tokyo University. After 10 years of studying and living in Japan he then decided to return to Vietnam.

In 2006, he started his firm, Vo Trong Nghia Architects in Ho Chi Minh City, Vietnam. The early projects were characterized by bamboo structures. From the first bamboo project, Wind and Water Café, he developed his bamboo construction team through his bamboo buildings. The design approach combined with bamboo treatment and construction brought him global awards and recognition.

Several years later, he also started green projects which utilize not only landscape plants, but started to design green façades and roofs which work as a bio-skin for buildings. His aim is to rehabilitate Vietnamese urban conditions with greenery in order to reconnect humans back to nature. "House for Trees" is a project that is now a series of projects working to bring more and more greenery to Vietnamese cities and aiming to solve the problem of urban pollution.

Nghia received international prizes and honors including but not limited to: AR House Award, six ARCASIA awards (among them gold medal and Building of the Year), and FuturArc Green Leadership Award, seven World Architecture Festival awards and the Prince Clause Awards in 2016. He also was selected as one of 2014's Young Global Leaders by the World Economic Forum.

Besides running his architectural practice, Nghia has continued to be involved in architecture at a grass roots level by teaching at the Singapore University of Technology & Design in 2015 and Hiroshima Institute of Technology in 2016.

Vo Trong Nghia is a registered architect in Vietnam.

Qualifications Registered Architect in Vietnam / 2005
ME Civil Engineering / 2004 / Tokyo University, Japan

Experience Representative - Vo Trong Nghia Architects, Vietnam / since 2006
Visiting Professor, Hiroshima Institute of Technology, Japan / 2016
Visiting Professor, Singapore University of Technology & Design, Singapore / 2015

Main Lectures

2020
Danang University of Science and Technology, Danang 12/11/2020
The University of British Columbia Online Lecture, Vancouver 19/10/2020
Archifest Virtual Conference, Singapore 28/09/2020/ Redefining the Landscape: Creating Innovative Architectural Concepts Through Buddhist Meditation and the 5 Precepts

2019
Cambridge University, London 01/11/2019 / Architecture and Meditation
Hanoi Architectural University, Vietnam 25/10/2019
Duy Tan University, Vietnam 10/06/2019
National Taiwan University of Science and Technology, Taipei 22/03/2019
Taichung City Assembly Hall, Taichung 22/03/2019
Tunghai University, Taichung 23/03/2019
Ming-Chuan University, Taichung 23/03/2019

2016
Innovative City Forum, Japan
Harvard University, USA

2015
Kyusyu Sangyo University, Japan
Hiroshima Institute of Technology, Japan
Gallery MA, Japan Columbia University, USA
Bartlett School of Architecture, UK
New Zealand Institute of Architects, NZ

2014
Singapore University of Technology & Design, Singapore
Institut Français d'Architecture, French Institute of Architecture, France
Australian Institute of Architects, Australia
Technische Universität Braunschweig, Germany

2013
Tsinghua University in Beijing, China University of Hong Kong, ChinaUniversity of Hong Kong, China

Main Jury World Architecture Festival, 2014
FuturArc Green leadership Award, 2014
Borderless competition, 2013

Associations Member of Vietnam Association of Architects (Vietnam)
Member of RIBA (UK)
Member of Japan Institute of Architects (Japan)

Awards

2006-2021

2021

- Yuanye Awards 11th / Winner/ Grand Prize/ Vedana Restaurant
- World Architecture Community Award 10+5+X 36th Cycle (WA Awards 10+5+X 36th Cycle)/ Winner/ Chicland Hotel/ Realised categories
- World Architecture Community Award 10+5+X 36th Cycle (WA Awards 10+5+X 36th Cycle)/ Winner/ Nocenco Cafe/ Realised categories
- World Architecture Community Award 10+5+X 36th Cycle (WA Awards 10+5+X 36th Cycle)/ Winner/ Farming Kindergarten/ Realised categories
- Wallpaper*Design Awards 2021/Winner/ Bat Trang house/The Best New Private house categories.

2020

- DFA Design for Asia Awards 2020 / Silver Award/ Stepping Park house/ Environmental Design/ Home & Residential Spaces category
- LOOP Design Awards 2020/ Winner / Viettel Academy/ Public building & Institutional category
- LOOP Design Awards 2020/ Honorable Mentions/ Atlas Hotel & Castaway Island Resort/ Hospitality & Hotels category
- Dezeen Awards 2020/ Winner / Thang House / Urban house of the year
- Dezeen Awards 2020/ Winner/ public vote/ Thang House / Urban house category
- Architecture Masterprize 2020 (MAP)/ Winner/ Chicland Hotel/Hospitality Architecture category
- Architecture Masterprize 2020 (MAP)/ Winner/Interior of the year/Nocenco Café/ Public Spaces category
- Dezeen Awards 2020/ longlist/ Chicland Hotel/ Hospitality building category
- Dezeen Awards 2020/ longlist/ Ha Long Villa/Urban house category
- FuturArc Green Leadership Award 2020, Winner/ Thang House/Residential- Individual houses category
- FuturArc Green Leadership Award 2020 / Merit/ Viettel Academy Educational Center/ Institutional category
- World Architecture Community Award 10+5+X 34th Cycle (WA Awards 10+5+X 34th Cycle)/ Winner/ / Ha house/ Realised categories

2019

- The ARCASIA awards, Bangladesh/The GOLD award/Castaway Island Rerort
- Dezeen Awards/ Winner/ Architect of the year/studios
- Dezeen Awards/ Winner/ Viettel Academy Educational Centre/ Civic-building
- Dezeen Awards/ Highly commended/ Stepping Park house/ Urban house
- Dezeen Awards/ Shortlist/ Ha house/ Urban house
- FuturArc Green Leadership award, Winner / Castaway Island Resort/Commercial
- FuturArc Green Leadership award, Merit / Stepping Park house/Residential- Individual houses.
- Green Good Design Awards / Bamboo Stalactite
- Green Good Design Awards / Nocenco cafe
- Green Good Design Awards / Phan Ke Binh house
- National Architecture Award [Vietnam], Silver / Viettel Academy Education Center
- National Architecture Award [Vietnam], Bronze / Castaway Island

2018
- World Architecture Festival (WAF) /Shortlist / Bamboo Stalactite / Display-Completed Buildings/ Small project of the year
- World Architecture Festival (WAF) /Shortlist / Nanoco Head Office/ Office-Future project
- World Architecture Festival (WAF) /Shortlist / Urban Farming Office/ Future project
- World Architecture Festival (WAF) /Shortlist / Chicland Hotel/ Leisure Led Development/ Future project
- World Architecture Festival (WAF) /Shortlist / Sheraton Phu Quoc/ Leisure Led Development/ Future project
- World Architecture Festival (WAF) / Inside Awards 2018/Shortlist / Nocenco Café/
- Green Good Design Awards / Bamboo House
- Green Good Design Awards / Binh House
- Green Good Design Awards / Dong Anh House
- Green Good Design Awards / Son La Ceremonial Dome
- Green Good Design Awards / Stacked Planters House
- International Architecture Awards (IAA) / Binh house
- International Architecture Awards (IAA) / The Lantern - Nanoco Gallery
- FuturArc Green Leadership award, Winner / Atlas Hotel
- FuturArc Green Leadership award, Winner / The Lantern - Nanoco Gallery
- FuturArc Green Leadership award, Merit / Binh House
- A'Design Awards / Building and Structure design/ Golden metal/ Farming Kindergarten.
- A'Design Awards / Building and Structure design/ Siliver metal/ House for Trees.

2017
- Ashui Awards / Green build of the year/ Binh house
- Word Architecture Festival (WAF)/ House- completed building Winner/ Binh House
- Word Architecture Festival (WAF)/ Office - Future project Winner/ Viettel Offsite Studio
- Word Architecture Festival (WAF)/ Education - Future project Hightly Commended/ Viettel Academy Educational Center.
- Zumtobel Group Award 2017/ Honorable mention/ Farming Kindergarten
- International Architecture Awards (IAA)/ Naman Conference Hall
- International Union of Architect Award (UIA)/ Honorable mention/ Farming Kindergaten
- The ARCASIA awards, Jaipur/The GOLD award/Son La restaurant
- The ARCASIA awards, Jaipur/The Honorable mention award/Roc Von restaurant and Son La restaurant
- The Green Good Design, USA/ Atlas Hotel
- The Green Good Design, USA/ Sen Village Community Centre
- The Green Good Design, USA/ Hoan House
- The Green Good Design, USA/ Roc Von Restaurant
- The Green Good Design, USA/ The Lantern

2016
- Prince Claus Awards/ Vo Trong Nghia
- DFA(Design for Asia) Award, Grand award / Naman Retreat
- ARCASIA 2016, Gold Medal/ Farming Kindergarten
- ARCASIA 2016, Honorable Mention/ S House 1
- ARCASIA 2016, Honorable Mention/ Factory Office Renovation
- ARCASIA 2016, Mention/ House for Trees
- BUILD Architecture Awards 2016, UK / Architectural Practice of the Year 2016 - Vietnam
- FuturArc Green Leadership award, Merit / Naman Retreat
- Green Good Design, USA/ Naman Retreat the Babylon

• Green Good Design, USA / Naman Retreat Hay Hay Restaurant and Bar
• Green Good Design, USA / Diamond Island Community Hall
• Green Good Design, USA / S House 3
• Green Good Design, USA / Vietnam Pavilion in EXPO Milano 2015
• Green Good Design, USA / FPT University Hanoi Administrative centre
• IAA (International Architecture Awards) / Diamond Island Community Hall
• IAA (International Architecture Awards) / Naman Hay Hay Restaurant
• IAA (International Architecture Awards) / Naman the Babylon
• Spec Go Green Awards/ Winning the first prize for Young Architect category/ Binh House
• Spec Go Green Awards/ Winning the third prize for Young Architect category/ The Lantern
• Vietnam Green Architecture Award of Vietnam Association of Architects, Gold / FPT University
• The Green Era Award / Sheraton Phu Quoc Resort

2015
• 8th artist in concrete awards ASIA FEST :2015-16 / winner of residential category
• 2015 Ashden International Awards / Runners-up / Greening cities with beautiful bamboo building
• ARCASIA award, Gold medal / Binh Thanh house
• ARCASIA award, Honorable Mention / Kontum Indochine Cafe
• AR school, finalist / Farming Kindergarten
• AR Emerging Architecture Award, Commended / Kontum Indochine Cafe
• Building of the year 2014 on Archdaily, Educational Architecture category winner / Farming Kindergarten
• DFA(Design for Asia) Award, Silver award / Farming Kindergarten
• FuturArc Green Leadership award, Winner / House for Trees
• FuturArc Green Leadership award, Citation / Son La Restaurant
• FuturArc Green Leadership award, Citation / Kontum Indochine Cafe
• FuturArc Green Leadership award, Citation / Green Renovation
• Green Good Design, USA/ Green Renovation
• Green Good Design, USA / House for Trees
• Green Good Design, USA / Factory Office Renovation
• Green Good Design, USA / S House 2
• Green Good Design, USA / Son La Restaurant
• Green Good Design, USA / Farming Kindergarten
• IAA (International Architecture Awards) 2015/ House for Trees
• IAA (International Architecture Awards) 2015 / Son La Restaurant
• IAA (International Architecture Awards) 2015 / Farming Kindergarten
• National Awards 2014 / Golden Award / Son La Restaurant
• National Awards 2014 / Golden Award / Binh Thanh house
• National Awards (Vietnam), Jury award / Farming Kindergarten
• National Awards (Vietnam), Team Award / Vo Trong Nghia Architects
• Vietnam House Award / Green Renovation

2014
• SPEC GO GREEN AWARDS / Winning prize for Young Architect category / FPT University
• AR Emerging Architecture Awards / highly commend / Farming Kindergarten
• IAA (International Architecture Awards) / Kontum Indochine Cafe
• AR House Award / winner / House for trees
• ARCASIA 2014 / Building of the year / Dailai Bamboo Complex
• ARCASIA 2014 / Gold Medal / Dailai Bamboo Complex

- Building of the year 2013 on Archdaily, Houses category winner / Binh Thanh House
- Green Good Design, USA/ Kontum Indochine Cafe
- Green Good Design, USA / Stone House
- Green Good Design, USA / Binh Thanh House
- Green Good Design, USA / Dailai Conference Hall
- Vietnam Green Architecture Award of Vietnam Association of Architects, Gold / Stone House
- Vietnam Green Architecture Award of Vietnam Association of Architects, Gold / Kontum Indochine Wedding Restaurant
- Wallpaper* Design Awards, Best private house category, shortlisted / Binh Thanh House
- World Architecture Festival (WAF), House category winner / House for trees
- World Architecture Festival (WAF), Hotel&Leisure category winner / Son La Restaurant
- World Architecture Festival (WAF), Future Projects Education category winner / FPT Technology University
- World Architecture Festival (WAF) /Shortlist / Farming Kindergarten
- World Architecture Festival (WAF) /Shortlist / Factory Office Renovation
- World Architecture Festival (WAF) /Shortlist / The Cam An Communal House
- Young Global Leader 2014 / World Economic Forum

2013

- ARCASIA 2013 (honorable mention) / Stacking Green
- Architect of the Year 2012, Vietnam / Vo Trong Nghia
- Building of the year 2012 on Archdaily, House category winner / Stacking Green
- FuturArc Prize (Professional category 2nd Prize), Singapore / Farming Kindergarten
- FuturArc Green Leadership award (Citation), Singapore / Dai Lai Conference Hall
- FuturArc Green Leadership Award (Citation), Singapore/ Low Cost House
- IAA (International Architecture Awards) / Dailai Conference Hall
- National Architecture Awards 2012, Vietnam/ 2nd Prize / Dai Lai Conference Hall
- National Architecture Awards 2012, Vietnam/ Citation / Stone House
- National Architecture Awards 2012, Vietnam/ Citation / Binh Duong school
- National Architecture Awards 2012, Vietnam/ Team Award / Vo Trong Nghia Architects
- World Architecture Festival (WAF) 2013/ Highly commended / Kontum Indochine Café
- World Architecture Festival (WAF) 2013/ Shortlist of Villa category / STONE HOUSE
- World Architecture Festival (WAF) 2013/ Shortlist of Civic and community category / DAI LAI CONFERENCE HALL
- World Architecture Festival (WAF) 2013/ Shortlist of Hotel, leisure category / KONTUM INDOCHINE CAFÉ

2012

- Architectural Record Award / Design Vanguard 2012
- FuturArc Green Leadership award (Winner), Singapore / Bamboo Wing
- FuturArc Green Leadership award (Citation), Singapore / Binh Duong School
- FuturArc Green Leadership award (Citation), Singapore / Stacking Green
- Green Good Design, USA / Stacking Green
- Green Good Design, USA / Bamboo Wing
- IAA (International Architecture Awards) / Stacking Green
- The WAN Awards 21 for 21, UK / Vo Trong Nghia Architects
- Vietnam Green Architecture Award of Vietnam Association of Architects/ Bamboo Wing
- Vietnam Green Architecture Award of Vietnam Association of Architects, Vietnam / Binh Duong School
- World Architecture Festival (WAF), School category winner / Binh Duong School
- World Architecture Festival (WAF), House category winner / Stacking Green